W9-CDD-629

# The Wonderful World of Birds and their Behaviour

# The Wonderful World of Birds
## and their Behaviour

Donald Broom

Hamlyn
London · New York · Sydney · Toronto

# Acknowledgements

**Aquila Photographics:** Horace Kinloch 7B; A. J. Moffat 30T; D. D. Richards 49. **Ardea Photographics:** J. A. Bailey 33T, 58T, 77; Hans and Judy Beste 41B, 72T; G. T. Brockhuysen 46T; Elizabeth Burgess 27T; John S. Dunning 36, 38, 39, 40; M. D. England 28B, 44B, 92B; Kenneth W. Fink 22, 24, 43, 59BR, 97; M. E. J. Gore 78B; John Gooders 92T; Clem Haagner *front jacket*, 91B; C. R. Knights 6; A. K. E. Lindau 62T; E. McNamara 48T; E. Mickleburgh 8; P. Morris 53, 88; Sid Roberts 27B, 47; Robert T. Smith 63B; Peter Steyn 29B; W. R. Taylor 37, 78T; Richard Vaughan 51B, 56L; Tom Willock 19. **Don Broom:** 23, 28T, 31B, 34BL, 51T, 73B, 83B. **N. Chaffer:** 75T. **Bruce Coleman Limited:** Bob Calham 33B; Peter Hinchliffe 69BR; Peter Jackson 79B. **Peter Eastman:** 83T. **M. P. L. Fogden:** 9. **Crawford H. Greenwalt:** 15. **Eric Hosking:** 7T, 14T, 30B, 32, 45, 61, 70, 71T, 74T, 80B, 95B. **Jacana:** A. Lagurgue 64; F. Varin 81T. **Frank Lane:** Ronald Austing 14B, 60, 76B, 91; Lynwood Chace 59T; Arthur Christiansen 67, 68B, 75B; Tom and Pam Gardner 29T; H. V. Lacey 90; Frank W. Lane 91T; W. T. Miller 46B, 89; G. T. H. Moon 68T; Georg Nostrand 62B; W. W. Roberts *back jacket flap*; Len Robinson 25; Ronald Thompson 26; Dr H. Wirth 31T; B. R. Young 11B; Dieter Zingel 13; Dr T. Zyrva 34–35. **S. D. MacDonald:** 71B. **Natural History Photographic Agency:** Anthony Bannister 79T; Frank V. Blackburn *title page*, 56R, 69T; C. Cameron 54; D. N. Dalton 57B, 69BL; Stephen Dalton *front jacket flap, back jacket*, 57T, 80T; J. Good 21; F. Greenway 81B; Brian Hawkes 7L, 84; E. A. Janes 11T, 76T, 95T; J. Jeffery 63T, 85; Peter Johnson 12, 17, 58B, 74B; M. Morcombe *endpapers*, 44T; K. B. Newman 55B, 65; P. Scott 72B; J. Tallon 94. **Colin Osman:** 87. **Tierbilder Okapia:** 73T. **Wildlife Studies Limited:** 10, 16, 20, 41T, 48B, 52, 55T, 59BL.

Published by the Hamlyn Publishing Group Limited
London · New York · Sydney · Toronto
Astronaut House, Feltham, Middlesex, England
Copyright © The Hamlyn Publishing Group Limited 1977

ISBN 0 600 31391 3

Filmset in England by Filmtype Services Limited,
Scarborough, Yorkshire
Set in 12 pt Bembo

Printed in Spain by Mateu Cromo, S. A. Pinto (Madrid)

# Contents

# Why are birds successful?

Birds can feed and breed in polar conditions. These Little Auks *(Plautus alle)* breed in colonies which may number over a million on islands in the arctic. They feed principally on crustaceans from the rich plankton of the polar seas.

Birds are a conspicuous part of man's surroundings in all parts of the world. Provided that there is an adequate food source available, it is likely that a bird will be there to exploit it. Birds can survive in a wide range of climates on land and on water because, with their basic body form, physiology and behaviour, they can adapt to most of the stresses which are imposed upon them by their physical surroundings. These adaptations will be discussed later in this chapter and in subsequent chapters.

The abundance of birds is further evidence for their success. Their numbers are most obvious when roosting flocks, breeding colonies or migrating birds are seen, but a census of singing male birds in mixed, temperate woodland reveals the presence of impressive numbers of birds. Tropical populations are even higher. Census work can be quite accurate for some species of birds which are easy to see or which congregate in a small number of places at certain times of year, but estimates of the total number of birds in a country or in the world can be no more than vague approximations. Nevertheless, Fisher's estimates of 100,000 million for the world population of birds and 120 million for the breeding population of Britain, and Peterson's estimate of 7,600 million birds breeding in the USA are generally considered to be realistic. The estimate for Britain is based upon surveys of different habitats and knowledge of the total areas occupied by such habitats in the country. The density of birds breeding in orchards or gardens was 750 per 10 hectares (25 acres) but that in fields of cereals or in coniferous woodland was only 50 per 10 hectares. Estimates for coniferous woodland in other places have varied from 30 per 10 hectares in northern Finland to 190 per 10 hectares in West Virginia. The highest population density estimates are 1,000 per 10 hectares for tropical grassland in Tanzania and 1,400 per 10 hectares for bird sanctuaries, which include a lot of woodland, in Germany and England. There have, however, been no satisfactory attempts to estimate numbers in what is perhaps the most impressive ornithological habitat; mixed regenerating forest in South America. Any ornithologist who can recognize bird songs can carry out a census of an area by visiting it on several occasions at different times of day during the breeding season and plotting on a map the song posts of all singing male birds.

Right
The Sahara Desert is very inhospitable for animals, but birds can live in all but the driest parts. This Deserk Lark *(Ammomanes deserti)* is sandy coloured so that it is difficult to detect against its background.

Below
Alpine Choughs *(Pyrrhocorax graculus)* have been observed just below the summit of Mount Everest at an altitude of over 8,000 m (26,400 ft).

Right
The Starling *(Sturnus vulgaris)* successfully exploits the habitats produced by man. It feeds in our agricultural areas and in our gardens. Large roosts are often formed in the centres of cities. The range of the Starling has expanded greatly in recent years, particularly in North America, and now covers a large proportion of the temperate regions of the Northern Hemisphere. It may well be the commonest wild bird.

Wilson's Storm Petrel *(Oceanites oceanicus)* is little larger than a sparrow, but it is an oceanic bird which comes to land for breeding only. The large numbers of these birds which are sometimes seen, and the vast areas of ocean over which the species can be found, have led to suggestions that the world population may be as great as those of the commonest land birds.

The most abundant species of bird in the world is almost certainly the Domestic Fowl *(Gallus gallus)* but there is some doubt as to which is the commonest wild bird. Species whose numbers have been estimated at well over 100 million include the Starling *(Sturnus vulgaris)*, the Quelea or Red-billed Dioch *(Quelea quelea)*, and Wilson's Petrel *(Oceanites oceanicus)*.

These population estimates are impressive, but it is difficult to understand their biological significance unless they can be related to estimates for other sorts of animals. Most animals are much less obvious than birds, however, so that any census of a large area is subject to considerable inaccuracy. It is therefore essential for the ecologist who wishes to assess the relative contributions of different animal and plant species in a community, to choose a small sample area for detailed study. If an area of heather moorland were studied, for example, the weights of heather, grasses, sedges, other plants, soil invertebrates, plant-eating insects, grouse, pipits, larks and hares could be estimated. Then, taking into account the lifespan of the animals and plants and various other factors, the contribution of each species to the total energy utilization in the community could be calculated.

It is only by means of such studies that the ecologist can assess the relative importance of different organisms. Those animals which feed directly on plants will always make the greatest contribution to energy turnover in the community. Most birds are predators, especially predators of insects, so the most interesting figure which the ecologist can provide is the impact of bird species on particular prey populations. One example of such a study is the survey of the ecology of the Ythan estuary near Aberdeen in Scotland. One of the principal sources of food for larger animals in the estuary is the mussel. This shellfish occurs in very large numbers and is eaten by man, fish, Eider Ducks *(Somateria mollissima)*, Oystercatchers *(Haematopus ostralegus)*, and gulls. In one year 54 per cent of all mussel production was consumed by predators. Eiders took 21 per cent, gulls took 8 per cent and Oystercatchers took 7 per cent. This sort of work is very difficult to do and most communities are even harder to study than the estuary, but it is now possible to provide some figures to support the statement that birds are successful members of most terrestrial and shallow-water communities.

# Feathers

Among the adaptations of birds which have contributed to their success are flight; warm blood and other physiological specializations; well-developed sense organs; a brain which allows precise analysis of sensory information and the control of complex behaviour; a sophisticated sound-producing mechanism; and a means of body coloration suitable for camouflage in some species and advertisement in others. One feature of avian anatomy, the possession of feathers, contributes to three of these adaptations. All birds have feathers but no other animal has them. The feather is formed from the cells of the skin in a manner which is similar to the formation of scales on the legs of birds or on the bodies of reptiles. The feather grows from the base inside a sheath. The cells become full of a hard protein called keratin and die. Only those cells joined together by the keratin will persist so that when the sheath splits, the characteristic structure of the feather will be visible. Although full-grown feathers are composed of dead tissue, the regions of the skin from which they grew are still alive and can later produce other feathers.

The first feathers which grow on young birds are short, fluffy down feathers which serve to insulate the body against the cold and also afford some protection against mechanical injury. It is generally thought that these were the original functions of feathers and that their uses for flight and for body coloration were later developments in the evolution of birds. The enormous advantage conferred upon birds by their ability to fly must be due especially to their possession of feathers. Birds' flight feathers have sufficient strength and firmness to remain quite rigid when moved through the air despite being anchored only at the base. They are constructed in such a way as to retain the links between barbs when

The Quelea *(Quelea quelea)* from Africa, is another contender for the title of commonest wild bird. There is little doubt that it is the most important bird pest in the world, and its numbers can now be estimated quite accurately due to extensive study in many African countries.

Right
Most birds will bathe in water and clean their feathers if the opportunity arises. When this Starling *(Sturnus vulgaris)* has wetted all its feathers, it will leave the water and vigorously ruffle them in order to shake off water droplets before flying away.

they move and to push most of the air before them rather than to allow it to pass through the feather. Despite this property of acting like a strong, supported membrane they are very light, and the wings are constructed so that they can be folded away easily when not in use.

Feathers do not break easily but they can become abraded if frequently brushed against hard surfaces and they are not as efficient for flight or for insulation if allowed to remain wet or dirty. Habits which help to keep feathers in good condition have been favoured by natural selection, and birds spend a significant proportion of their lives engaging in feather maintenance activities. The general behaviour of birds is adapted so that feathers are not abraded unnecessarily. Birds refrain from flapping their wings against the ground, trees or rocks, whenever possible, and they do not creep through crevices if their feathers would be damaged by doing so. Dirt on the wings is removed by bathing in water or in dust and any water or dust is subsequently removed by flapping the wings, ruffling the feathers and by preening movements. Preening involves passing the bill, and sometimes the feet, through the feathers in such a way as to remove any material adhering to the feathers and to smooth and rearrange the feathers. The actions are often carried out after bathing or in association with the application of oil from the preen gland (otherwise known as the oil gland). This gland is found in most birds and is situated on the rump above the base of the tail. It produces an oily secretion and is most active in aquatic birds. The secretion probably helps to keep feathers supple so that they do not break, and increases their waterproofing qualities.

Despite their resistance to wear, feathers must be renewed at intervals. The moulting process, during which old feathers fall out and are replaced by new ones growing in the same place, also allows birds to change from juvenile to adult plumage or from non-breeding to breeding plumage. Feathers usually fall out and are replaced in a regular sequence starting with the wings and tail and followed by the body feathers from rump to head. The timing of moult depends upon temperature and level of nutrition, the activation of the process being due to the functioning of the thyroid gland. The main flight feathers moult from the centre of the wing towards the tip (primary feathers) and towards the body (most of the secondary feathers) at the same time. The rate at which flight feathers are lost varies greatly according to the way of life of the bird. Those birds which need to

Above
Preening behaviour, as shown by this Meadow Pipit *(Anthus pratensis)*, is very important for feather maintenance and may continue for many minutes. The bird may engage in an elaborate series of actions in which successive feathers in different regions of the body are systematically cleaned by nibbling movements or by drawing feathers through the bill. On other occasions, preening movements may be merely brief flicks of the bill which smooth a feather into place or remove a single particle of dirt.

Right
One curious activity shown by many passerine (perching) birds which may be a form of feather maintenance is 'anting'. The bird either grasps an ant in its bill and applies it to feather tracts on the extended wing, as does the Blue Jay *(Cyanocitta cristata)* of North America, or spreads out its wings on the ground by an ants' nest, thus allowing ants to swarm on to its plumage. It is assumed that the acids or other substances produced by the ants must benefit the birds in some way.

The male hornbill (Bucerotidae) walls up the female in her nest hole and feeds her through a tiny hole. While incubating her eggs, the female hornbill does not need to fly, so she can safely moult almost all her feathers at once. When male hornbills moult, they always retain enough feathers to allow efficient flight.

retain efficient powers of flight seldom have more than 20 per cent of their feathers missing at any one time but water birds can still find food and escape adequately from predators when flightless, thus ducks can moult all their flight feathers at once. Moulting birds are sometimes obvious because they shed feathers or because they have symmetrical gaps in their flight feathers. All birds moult at least once a year and many species moult twice a year. Since flying efficiency is impaired during moult, and energy is required to produce new feathers, moulting is not normally carried out at the times when the bird must be at its most efficient, for instance, when breeding, migrating, or in difficult winter conditions. Moult may sometimes start and then stop again before completion if adverse conditions occur. The brightly coloured males of many duck species characteristically have a post-breeding moult when they assume a dull coloration before losing their flight feathers. They are thus rendered less conspicuous during the time that their new flight feathers are growing.

## Flight

The wings of birds include all the bones which make up the fore-limbs of other vertebrates except for a reduction in the number of 'hand' bones. The humerus, which is joined to the body at the shoulder, and the radius and ulna bones which make up the forearm in man, bear the secondary flight feathers. The primary flight feathers are attached to the second digit and to a fused carpal and metacarpal bone which joins the digit to the forearm. The first digit bears the small alula or bastard wing, the third digit is vestigial and the fourth and fifth digits are missing. The other very obvious adaptation of the skeleton for flight is the enormous breast-bone to which the pectoralis muscles which pull the wing down, and the smaller supracoracoideus muscles which pull the wing up, are attached. Other adaptations of skeletal structure for flight include the fusion of bones in the pectoral and pelvic girdles and in the vertebral column, which results in the increased rigidity of the body needed for flight. Weight is reduced by the loss of tail vertebrae, teeth and heavy jaws and by the possession of hollow bones. The bones are hollow tubes with cross struts, a structure which modern engineers have found to confer maximum strength for a given weight of material. Some bones are also strengthened by pneumatization, their cavities being filled by extensions of the bird's elaborate system of air sacs. The characteristic shape of any bird is due in part to the air sacs which lie under the skin. These inflated sacs reduce the average density of the bird, although not the weight of course, and improve its streamlining.

In addition to the possession of strong light feathers and bones, the weight of birds is also reduced by rapid digestive processes and the avoidance of dense food; the reduction of all glands and the reproductive organs to the minimum functional size; reproduction by means of eggs rather than by the production of live young; and the excretion of light crystalline uric acid so that large volumes of water are not needed for urination. In order to fly, birds need a great deal of energy which must be obtained from their food. The initiation of movement is facilitated by the fact that they are warm-blooded and by the high level of glucose in the blood. A plentiful supply of oxygen is needed by the muscles and this is initially taken up by the blood in the lungs which form part of a through-flow system of air movement which involves the air sacs. The blood is transported to all parts of the body by an efficient four-chambered heart and circulatory system.

The bird's wing is convex on the upper surface and concave on the lower surface. It is rounded at the leading edge and tapers to the trailing edge, and is therefore an aerofoil. This shape has been found by aircraft designers to be the most efficient for flight. When such a wing moves through the air, leading edge first and parallel to the ground, the air has further to travel over the upper surface than over the lower surface and the net effect is a force which lifts the bird. This lift will disappear if the wing stops moving forwards, but most birds achieve forward momentum by flapping. Gliding birds will always lose height or speed in still air but there is great variation in the distance which a bird can glide from a given height according to the shape of the wings and the weight of the bird.

The Swift *(Apus apus)* spends most of its life on the wing. Its small feet are just strong enough for it to cling to a vertical cliff or to shuffle along the ground to its nest.

Right
Albatrosses, like this Waved Albatross *(Diomedea irrorata)* spend much of their lives on the wing. Their wings are very long and narrow, so that they can glide steeply and fast. They utilize the up-currents of air over a wave to gain height, and then glide down to the top of the next wave.

Above
The Turkey Vulture *(Cathartes aura)* from North America glides with its wings in a characteristic dihedral position so that the tips point diagonally upwards. The effects of the weight of birds on their gliding flight is very noticeable among vultures which have eaten large meals. They glide faster and are not able to glide as far. This sometimes leads to difficulties for birds living in areas where up-currents of air are widely separated from one another.

The shape of the wings can be discussed in terms of their aspect ratio, which is the square of the wing span divided by the wing area. Vultures have very broad wings so the aspect ratio is quite low and they can glide slowly at a very shallow angle. Albatrosses have very long wings so the aspect ratio is high and they are adapted to glide steeply and fast. All gliding birds must find rising air in order to keep flying. Land birds find up-currents near cliffs, over woods or in deserts, while oceanic birds utilize small rising currents over the tops of waves. All the gliders fall in relation to the air around them but if that air is itself rising they can increase their distance from the ground. Slow gliders deal with the problem of stalling by using feather tips and the bastard wing as 'slots' which prevent turbulent air flowing over the upper surface of the wing.

Birds take off from the ground by lifting the wings and then beating down with them parallel to the ground and jumping at the same time. During the upstroke, the wing is twisted so that it pushes much less air up than it will push down when it beats down again. This is the general principle of gaining height by flapping. Forward movement is achieved by holding the wing on the downstroke at such an angle that air is pushed downwards and backwards and on the upstroke at such an angle that air is pushed upwards and backwards. The downwards and upwards pushes cancel out and the net push on the air is backwards. In level flight the bird has to exert a slightly larger net downwards push to counteract the sinking effect due to its own weight but it is helped in this by the aerofoil shape of the wing. The shape of the wings also varies greatly among birds which use flapping flight most of the time. Short wings, which give a low aspect ratio, are best for very rapid

This Streamertail Hummingbird *(Trochilus polytmus)* is hovering in order to feed. Hummingbirds are the only birds which can hover in still air.

manoeuvring but are usually associated with a large weight per unit wing area (wing loading) so they must be flapped fast and much energy is needed to keep flying. Such wings are characteristic of game birds which do not need to make long flights but need to take off rapidly and vertically and to fly fast. Birds like swallows, which need to fly fast for long periods, have a much higher aspect ratio and a lower wing loading and the wings are more pointed. Improved manoeuvrability is produced by the use of a fairly long tail in such birds.

Hovering flight is possible for any bird in moving air for they merely need to fly into the air current at the same speed as it is blowing them backwards. This is the principle used by the European Kestrel *(Falco tinnunculus)* and various other birds of prey. Hovering flight in still air is possible only if the bird can flap its wings backwards with as much force as it flaps them forwards. In hummingbirds the supracoracoideus muscles, which pull the wing upwards in level flight, are almost the same size as the pectoralis muscles which beat the wing downwards. When hovering, hummingbirds hold their bodies almost vertical and the wings push on the air alternately down and backwards and down and forwards. The forwards and backwards pushes cancel out, and the downwards push is just sufficient to prevent the bird from falling.

# How birds maintain their body temperature

Birds need to maintain their body temperature at a high level in order that they can use their muscles efficiently at short notice if danger threatens. If they were cold blooded they would need a warming-up period before they could engage in normal activities and they would lose some of the most efficient aspects of body biochemistry which depend upon a relatively stable, high body temperature. The mechanisms of temperature regulation are partly physiological and partly behavioural.

Feathers provide excellent insulation from the cold, provided that they are not wetted so that the layer of air next to the skin is lost. Heat is provided during the part of the day when the bird is active by the chemical reactions which occur when food is broken down to produce energy, especially by the muscles. Birds can, therefore, survive cold conditions during the daytime provided that they seek shelter when rain or snow is falling or when a strong cold wind is blowing. During the night, when they may not be able to find food, they must remain as still as possible in a sheltered place so that heat loss and energy consumption are minimized. The everyday behaviour of the birds is thus adapted to conserve energy in the cold conditions they may have to face.

Extreme heat would also be damaging to birds. The principal means of avoiding over-heating is to keep out of the direct rays of the sun. Studies of a variety of birds in hot deserts reveal that even species which normally sit on top of bushes or rocks, such as shrikes and wheatears, will sit inside bushes and in other shaded places during the hottest part of the day. A

Left
The insulating properties of feathers are well known because the down feathers which the duck Eider *(Somateria mollissima)* uses to line her nest are collected by man to fill sleeping bags, bed coverings and pillows.

Above
The constant flow of birds to isolated water sources in dry habitats shows the necessity of drinking, but birds need less water than most mammals and need to expend less energy and time getting to the water. These Namaqua Sandgrouse *(Pterocles namaqua)* regularly visit water-holes.

behavioural adaptation to combat cold which certain small birds share with bees and some mammals is to huddle together. Well-known examples are two species which occur in Europe and in North America. Both the Wren *(Troglodytes troglodytes)* and the Tree Creeper *(Certhia familiaris)*, which are called Winter Wren and Brown Creeper in North America, will congregate in holes during cold weather and, by huddling together in groups which often include more than ten individuals, they conserve body heat.

Water loss is also a problem for any animal, particularly one living in hot conditions. Water is lost during breathing, but birds do not sweat and they lose little water during excretion. Nevertheless, the avoidance of direct sunlight where possible will reduce water loss as well as prevent overheating. Birds obtain water from their food and supplement this by drinking. Species which eat very dry food such as seeds need to drink more than those feeding on fruit or other food sources which contain much water. Seabirds are surrounded by water, but it contains much more dissolved salt than their blood. Cormorants, petrels and penguins are, however, able to eat salty food and to drink the water because they have a salt secreting gland in the head which is able to remove very much more salt from the body fluids than the kidney can remove.

# Sense organs and the brain

Animals which move rapidly need efficient sense organs to provide information about their surroundings and also about the relative positions of the different parts of their bodies. Birds have particularly sophisticated eyes and ears and also well-developed sense organs in muscles and joints. Each sensory system includes the part of the brain which analyses the information transmitted to it along the nerve from the sensory receptor as well as the sense organ itself. Most birds find their food and detect predators principally by visual means. The eyes of birds are proportionally larger than those of most other animals. The Starling's *(Sturnus vulgaris)* eye is about 1·5 per cent of its body weight but that of man is only 0·1 per cent of body

Right
The eyes of birds are generally large in proportion to their body size and those of owls, such as this Saw-whet Owl *(Aegolius acadicus)* are particularly large. Due to the high density of sensory cells in the retina, owls can see and catch their prey in illuminations which are about one-fortieth of that which man would need to locate objects.

weight. Birds of prey and owls have even larger eyes, some of them larger than man's eyes despite man's very much greater body weight. Birds' eyes are different from those of any other animals in certain respects but they are also very variable. The pecten, a comb-like structure situated between the retina and the lens, is much more elaborate in birds than in reptiles and is absent in mammals. It is best developed in birds which hunt active prey and its function is probably to enhance movement detection. The retina of most birds has a very high density of light receptor cells. Birds of prey have more 'cones' per unit area of retina than any other animal and, like other birds, they have large numbers of coloured oil droplets above the cones. The cones are used for colour vision, which is similar to man's, and to allow the formation of precise images in light conditions. Owls and other nocturnal birds have very large numbers of 'rods' per unit area of retina, for rods are used in dim light. The result of these anatomical specializations is that birds have more efficient vision than other animals. Man's ability to distinguish from a distance objects which are close together is better than that of most animals, but there is no doubt that many birds have a visual acuity which is just as good, or better. It has been estimated that birds of prey may have acuity which is as much as eight times better. Further points about visual ability are mentioned in the chapter on migration and navigation.

The widespread use of songs and calls by birds, as well as the importance of detecting the approach of predators which cannot be seen, has resulted in natural selection for an efficient sense of hearing in birds. The ears of birds respond to sounds at a wide range of frequencies. The range and optimum frequency sensitivity varies according to the way of life of the bird but most are similar to, though slightly narrower than, man's frequency range. The assessment of the range of frequencies which birds can hear is complicated by the presence of vibration receptors in other parts of the body, especially the legs, which respond to low frequencies. It has recently been discovered that birds flying high over breaking waves can detect the low frequency sound which they produce. Birds cannot detect ultrasonic sounds in the 40–80 kHz range such as those used by rodents for communication or by bats for navigation. It seems likely that the ability of birds to detect low intensity sounds may be better than that of man and it is certain that birds are better at distinguishing between sounds which occur close together in time. The patterns of bird songs which are so rapid that we hear them as a buzz can be imitated by other birds, so they must hear the individual notes which we cannot detect unless we use electrical aids.

The sense of touch and those senses involved in the control of movement are efficient in birds but are not remarkably different in mechanism from those of other vertebrates. The sense of smell was long thought to be unimportant in most birds, but the sense organs and the part of the brain which deals with information from them are well developed in kiwis *(Apteryx)*, and in certain geese, petrels and vultures. It seems likely that members of the petrel order, which includes albatrosses and shearwaters, may detect food by smelling the wind which blows over the ocean surface and that vultures might occasionally smell carcasses which they do not see. Further evidence about the use of the sense of smell and the possible existence of other senses is mentioned in the chapter on migration and navigation.

The brain of birds evolved to its present considerable complexity due, in part, to the need for a mechanism to control the complex muscular activity needed to fly and the need for analysis of information from sense organs. The control of movement is possible only if precise sensory information is available so that the two systems are interlinked. Much of the sensory analysis is carried out by the cerebral hemispheres and these are large in birds, although not as large as in some mammals. The largest region within the hemisphere is the striatum in birds but the pallium is largest in mammals. The control of movement takes place in the cerebellum and this is proportionally larger in birds than it is in mammals. The very large size of the brain of birds in comparison with their reptilian ancestors is also due to the

Of all the characteristics and attributes of birds, the one which has the greatest impact on many people is their ability to sing. It seems likely that our music and our ideas about aesthetics have been much influenced in their development by the songs of birds. Most people are affected by hearing the sounds of bird song and would consider bird song to be an important aspect of their environment. The Nightingale *(Luscinia megarhynchos)*, which is seen singing here, is one of the best-loved songsters of northern Europe.

need to control other aspects of their behaviour in addition to their locomotor movements. Their elaborate feeding behaviour and social behaviour are certainly reasons for their success in the animal world and these topics are discussed further in subsequent chapters. The behaviour would not be possible without the brain mechanisms to control it.

## Song

The ability to produce songs and calls has played no small part in determining the success of birds. The song is important in proclaiming territory and is, therefore, a very valuable means of ensuring that the maximum number of the most successful individuals find mates and that breeding pairs are adequately separated from one another. Bird calls are even more important as a means of warning other individuals of the approach of predators and as a method of preserving contact between the members of a group. Social behaviour with all its advantages would scarcely be possible without this important communication mechanism. The noises made by birds are produced by blowing air through a specialized organ called the syrinx. This is composed of a set of elastic vibrating membranes, whose tension can be altered by the action of a series of muscles, which is situated in a resonating chamber. The sequences of sounds can be controlled very precisely and several notes can be produced at the same time by some species.

Bird songs are extremely varied, since they are used by the birds to distinguish between their own and other species and, in some cases, between individuals. Bird calls are much more uniform within species, since the birds need to be able to interpret the calls of all other individuals. When a flock of finches is moving from one part of a feeding area to another, they need to retain contact with one another, so all the individuals give a contact call which is characteristic of the species. Another situation where contact calls are valuable is in migrating flocks. The high pitched 'seep' call of migrating Redwings *(Turdus iliacus)* will be familiar to many in northern Europe who have listened for these birds at night in late autumn or spring. The uniformity of calls extends across species as well as within species in certain situations. The alarm call given when a ground predator approaches varies considerably from one species to another, for the optimum avoiding action will be very variable, but the alarm call which is produced when an avian predator appears is remarkably constant for a wide range of small bird species. If a hawk or falcon is seen by members of the Paridae (tits and chickadees), Fringillidae (finches, buntings and sparrows) or Turdinae (thrushes) the bird gives a call in which the high pitch does not change and which starts and ends gradually rather than abruptly. All other small birds which hear this call will remain still if they are already hidden, or fly rapidly to the nearest cover if they are in the open. A call of this kind is particularly difficult for predators to locate and all the different potential prey species benefit one another by producing and responding to the same call.

# Avoiding predation

Despite the fact that the majority of birds eat other animals and are therefore predators, all but the very largest are subject to predation by large birds of prey and by predatory mammals. It is, therefore, very important to them to have anatomical and behavioural adaptations which reduce the chances of any individual being eaten. The two principal ways in which a small bird can avoid predators are to be very good at hiding or to be very good at escaping. Escape by flying is the most obvious method but it is inadequate if the predator is another bird and is also inadequate against ground predators because most birds must stop flying in order to feed, preen or rest, and they are then vulnerable to fast-running predators. When a bird is resting it must either conceal itself or it must rest in a place where an approaching predator would be obvious enough to allow it time to escape. Concealment involves selection of particular resting places in which the bird is actually invisible to predators or is effectively invisible because of its coloration. Diurnal birds, those active during the daytime, usually spend the night roosting off the ground, where they are relatively safe from ground predators. The Pheasant *(Phasianus colchicus)* will spend the whole day feeding on the ground but will fly up into a tree at night, where it will be safe from foxes. Should a predatory mammal which can climb try to catch it, the climber will usually make sufficient noise to alarm the bird. The safer the hiding place, the less alert for danger the bird needs to be.

Most birds which rest on the ground during the day time are cryptically coloured. They may resemble wood and leaf litter like many game birds (Galliformes),

The Woodcock *(Scolopax rusticola)* spends most of the day sitting on the ground. It exploits the plumage coloration by finding a place to rest where its markings will blend with the background and by keeping very still when danger threatens.

and nightjars (Caprimulgidae); or they resemble earth or lumps of vegetation like the Stone Curlew *(Burhinus oedicnemus)*; or they are streaked so as to resemble their reedy habitat like bitterns *(Botaurus)*; or they are sandy coloured like the Desert Lark *(Ammomanes deserti)*; or white like many birds living in snowy regions, for instance the Ptarmigan *(Lagopus mutus)*. The plumage adaptations are always associated with behavioural adaptations. The bird chooses a suitable resting place in which it will benefit from its cryptic coloration and it orients itself and remains still when in that place. A bittern is conspicuous when standing in open water or in a field and is quite obvious, even in a reed-bed if it stands so that its streaks run at right angles to the reeds or if it moves around. The advantages which prey species obtain from concealing coloration may also be advantageous to large predators which, if camouflaged, are not noticed by their prey until they start the pursuit. An example is the Snowy Owl *(Nyctea scandiaca)*.

Successful escape from a predator depends upon detecting the predator in time to initiate the escape response. Birds which are very susceptible to predation, therefore, spend a large amount of time maintaining alertness when they are in hazardous situations, so the eyes are set on the sides of the head, thus giving all-round vision. As will be mentioned in the chapter on flocking, roosting and territory, the risk is reduced in social species because there are more pairs of ears and eyes on the look out for danger. Every individual bird must find a balance between time spent feeding, or engaging in some other essential activity, and time spent looking out for predators. The response when a predator is detected may be to freeze, or to crouch to the ground and freeze. This response is shown by almost all birds but is most frequent in cryptically coloured birds. When immobile, the bird must continue to watch the predator, if possible, just in case it has been seen. Escape movements may involve flight, but some birds usually run away from predators, for instance, the bustards (Otidae) and, of course, the Ostrich *(Struthio camelus)* and other large flightless birds, while others take to water or dive from the surface when attacked, for instance, grebes (Podicipitidae). The escape reaction shown when a slow-moving ground predator such as man approaches will be slower than that shown when a fox appears, and the response to a hawk or falcon is likely to be very fast and directed towards dense cover rather than a vantage point from which a ground predator could safely be observed. Special anti-predator behaviour is shown during nesting and this will be mentioned in the chapter on courtship and nesting.

The Ptarmigan or Rock Ptarmigan *(Lagopus mutus)* lives in rocky places where its coloration matches that of the rocks. When the bird remains still it is very difficult to see. Before winter starts they moult and the new feathers are white, so that the bird is concealed in snowy conditions.

# Evolution and the principal families

Comparative anatomists have long known that, while birds share many features with mammals, they are fundamentally different in other respects. Both groups of animals have the same sort of general body plan as reptiles. Even without any real evidence of ancestral forms, it would be concluded that they evolved independently from reptilian ancestors. This conclusion was considerably reinforced by studies of fossil reptiles, for there are some fossils which are undoubtedly reptiles, but which are more like mammals than are any living reptiles, and there are other fossils which are a little more like birds than are any living reptiles. The situation was further clarified by the discovery of fossil mammals and fossil birds with anatomies intermediate between those of living forms and those of fossil reptiles.

There are many similarities between birds and their nearest reptilian relatives, the dinosaurs and crocodiles. They have similar brain case, jaw and neck structures, and the legs, ribs and pelvis have features in common. Both have scales, both lay eggs and there are similarities between the brain, sense organs and blood of each. Birds are more obviously different from the other main lines of reptilian evolution which have given rise to turtles, lizards, snakes and mammals respectively. The order of reptiles which was ancestral to birds was the Thecodontia, whose members lived 200 million years ago during the Triassic period. Among the reptiles which evolved from the Thecodontia were the winged pterodactyls, although these were not immediate ancestors of birds. We still do not know for sure how the large step from a reptile to a warm-blooded, flying bird took place, but some help in understanding this problem was provided by the discovery of the fossil bird *Archaeopteryx* in Bavaria in 1861.

Birds have delicate skeletons and, because of their way of life, they are not very likely to become embedded in mud or other material and subsequently fossilized. This partly explains why so few bird fossils have been found, but the earliest birds were probably not very common, thus making it even less likely that their fossils could be found. This scarcity of bird fossils emphasizes the great fortune of palaeontologists in finding *Archaeopteryx* which lived about 150 million years ago in the late Jurassic period.

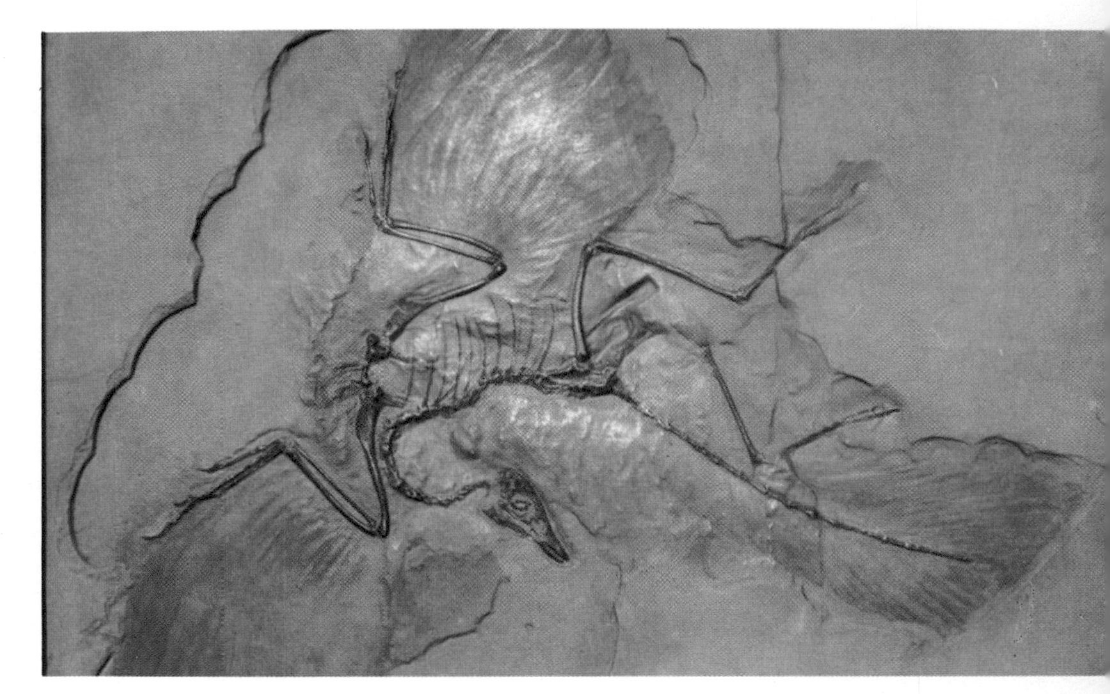

Three specimens of the fossil bird *Archaeopteryx* have been found in Germany, all in limestone beds. The animal had a dinosaur-like skull and toothed jaws, a long bony tail, clawed fingers and ribs in its abdomen. None of these characters is possessed by living birds, but the wings and tail of *Archaeopteryx* are very obviously feathered, and the beak, feet and wings are very bird-like.

Kiwis *(Apteryx)* have short legs, rounded bodies, hair-like feathers and a long flexible bill with which they probe in the earth for earthworms, insects and vegetable food. The nostrils are at the tip of the bill and the nasal organs are large so the birds may find food by means of smell.

Further bird fossils which resemble modern divers, grebes, cormorants and gulls in general body form are found in rocks laid down in the Cretaceous period of about 100 million years ago. It is not surprising that the marine birds should have become fossilized more readily than terrestrial birds and the considerable extent to which these Cretaceous fossils resemble modern birds makes it seem likely that a wide range of terrestrial forms might have existed at this time. Land birds were common enough fifty million years ago to include representatives of twenty-seven of the families of birds living today. The period of the greatest proliferation of birds was well after the demise of most dinosaurs, but was perhaps fifteen to twenty million years earlier than the peak of diversification among mammals. One consequence of this was the evolution of a number of types of ground-dwelling birds which could not compete adequately with the mammals and which became extinct again. This process has continued until recent times, for large flightless birds survived best on islands, which many mammals did not reach, but were finally made extinct by man. A recent example from the North Atlantic is the Great Auk *(Pinguinus impennis)*, while others include the Dodo *(Raphus cucullatus)* of Mauritius, moas of several species such as *Megalapteryx hectori* in New Zealand and elephant-birds *(Aepyornis* spp*)* in Madagascar, all of which survived until 300–400 years ago. It is generally considered that all modern flightless birds evolved from ancestors which could fly.

There is no doubt that, like all other animals including man, birds are still evolving today. Many species have become extinct and others have become very much more common. The extent to which any species can vary is exemplified by the great variety of different breeds of domestic birds which have been produced by the extreme selection pressures exerted by breeders choosing among naturally occurring genetic variants. The surroundings in which some birds live are changing rapidly and so some genetic variants will be more successful than others and will change the average characteristics of species. The effects of selection in the past are apparent when the diversity of birds in any area is considered. There is a variety of possible food sources, and the birds in any geographical region have become adapted to exploit them. In some regions a large variety of families of birds was originally present but in other places, especially on islands, a limited variety of forms has diversified to exploit a range of habitats. An extreme example of this occurred in the Galápagos Islands, which must originally have been colonized by some seed-eating finches from the mainland. After thousands of years they have evolved into species which are tit-like, warbler-like, nuthatch-like and parrot-like, as well as several which are still seed-eaters.

# The main groups of birds

Birds are all members of one class of animals whose scientific name is Aves. This class is divided into twenty-seven orders with living representatives. Most of these orders include several or many families. The largest order, the Passeriformes or perching birds, includes more than half of the known species of birds, while the smallest order, the Struthioniformes, includes only one species, the Ostrich *(Struthio camelus)*. All birds have a scientific name which is composed of two words, the first being the name of the genus and the second the name of the species. An example of a genus is *Falco,* the falcons, which includes the Peregrine Falcon *(Falco peregrinus)*, the Lanner Falcon *(Falco biarmicus),* the European Kestrel *(Falco tinnunculus)* and the American Kestrel *(Falco sparverius)*. Minor regional variations within a species are denoted by the use of a sub-specific name which is put after the specific name. Members of a species will breed together successfully and produce fertile offspring in the wild, while hybrids between species are very rare except between certain species in captivity. There are, however, species with an extended range which vary over that range so that the forms at the extreme ends are dissimilar from one another. Since these individuals would not normally meet, the question of whether they would breed successfully does not arise. It does seem likely, however, that many species now considered separate from one another have arisen from a single species by geographical isolation. Darwin's finches on the Galápagos Islands have already been mentioned. A different example is provided by the gulls of the Herring Gull group. A range of species exists in a band around the whole of the Northern Hemisphere. These may have spread out from an ancestor at one point. In Europe two species exist, the Herring Gull *(Larus argentatus)* and the Lesser Black-backed Gull *(Larus fuscus)*, which do not interbreed and which may be the results of the spread of this type of gull in two directions around the world.

The Emu *(Dromaius novaehollandiae)* is the second largest bird in the world. It is still widespread in parts of Australia, especially in Western Australia. The feathers are double due to the aftershaft, which is small in most bird feathers, equalling the main shaft in size. The barbs are not joined together, for the principal function of the feathers is insulation.

The following list of bird families is not exhaustive, but includes some mention of the largest and most remarkable families.

**Kiwis**

These flightless birds are not uncommon in some parts of New Zealand, although few New Zealanders have seen one. The three species of kiwi *(Apteryx)* range in size from that of a bantam to that of a large cockerel. The egg is very large and may be 15 per cent of the adult female's weight.

**Ostriches**

The single species in this order, *Struthio camelus,* is the largest living bird. An adult male may stand almost 2·5 m (over 8 ft) in height and may weigh up to 150 kg (330 lb). The legs are very large and the toes reduced to two in number. Escape from predators is effected by swift running across the open desert or grassy plains which Ostriches inhabit. The other two orders of large, fast-running flightless birds include the rheas of South America and the Emu and cassowaries of Australasia.

## Grebes

This order of birds includes twenty-one species. All spend most of the time swimming and diving for food, but they can fly rapidly if necessary. The breeding plumage of grebes and divers is often very striking, and their courtship displays are elaborate. One of the first detailed studies of animal courtship was that of J. Huxley on the Great Crested Grebe *(Podiceps cristatus)*.

## Penguins

The fact that penguins stand upright and are often dark on the back and white underneath has resulted in comparison, however fleeting, with people wearing black and white clothes. Their gait and apparent expression usually evoke mirth. If, however, they are seen swimming, the advantages of their body form become obvious. They swim by using their diminutive wings with a movement not unlike flying through the water. The largest number of individuals is found in Antarctica, where they may breed far from the sea in large 'rookeries', but some penguins live as far north as the Galápagos Islands on the Equator.

## Petrels

Most of the members of the order Procellariiformes, which includes petrels, shearwaters and albatrosses, are truly oceanic for all of their lives except at breeding times. The largest species, the Wandering Albatross *(Diomedea exulans)*, has a wing span of 3·5 m (11·5 ft) and, like the other large members of the order, it flies with little flapping, utilizing instead the upward currents of air over each wave.

## Frigate birds

These splendid birds, which are in the same order as the pelicans, cormorants, gannets and boobies are confined to tropical and subtropical regions. The largest species, such as the Magnificent Frigate Bird *(Fregata magnificens)*, have a wing span of just over 2 m (6·5 ft). They have a long hooked bill which they use to catch flying fish and other fish on the water surface, but their principal method of feeding is to chase smaller birds which have caught fish and force them to disgorge. The fast-flying manoeuvrable frigate bird then swoops down and catches the fish before it hits the water. These birds cannot land on the water surface, as their plumage is not adequately waterproofed.

## Ibises

Ibises have long bills and long legs like the spoonbills, which are members of the same family, and the herons, egrets and storks, which are in closely related families. One of the twenty-six species, the Sacred Ibis *(Threskiornis aethiopica)*, was revered by the ancient Egyptians, who painted it and mummified its

Left
Cormorants *(Phalacrocorax carbo)* are in the same order as gannets, boobies and frigate birds. Their long, slightly hooked bills are well adapted for grasping fish, which they catch underwater.

Above
The Roseate Spoonbill *(Ajaia ajaia)* has a large spatulate bill which it uses to catch crustaceans and small fish as well as to preen itself.

Right
The Black-throated Diver or Loon *(Gavia arctica)* is one of five species in an order of birds which are well adapted to an almost entirely aquatic life. The body is streamlined for underwater swimming and the lobed feet are set far back on the body.

Right
The Great Egret *(Egretta alba)* is one of the most widespread members of the heron family. In North and South America it is also called the Common or American Egret. It also lives in Africa, south-east Europe, Asia and Australasia. The spear-like bill is well adapted for fishing.

Below
Greater Flamingos *(Phoenicopterus ruber)* feed by holding their heads upside down. The bill is dipped in the water and swung from side to side so that food can be sieved out. Flamingos show some similarities with herons, but more with ducks, geese and swans.

Top right
Black Swans *(Cygnus atratus)* are Australian species, but are frequently kept in captivity as ornamental birds. Swans, geese and ducks are all members of a single family.

Bottom right
This Lanner Falcon *(Falco biarmicus)* has the hooked bill, strong talons and compact body typical of the more active birds of prey.

remains. The species still occurs over much of Africa, but it is no longer found in Egypt. Another well-known ibis is the Scarlet Ibis *(Eudocimus ruber)* which is the national bird of Trinidad. This brilliant scarlet bird feeds on small crabs in areas of mangroves, and its fame is due especially to its habit of roosting communally. Large flocks of these beautiful birds flying to roost are an unforgettable sight.

**Ducks**
The family Anatidae includes 145 species of ducks, geese and swans. The Mallard (*Anas platyrhynchos)*, is the commonly domesticated duck species and, together with its close relatives, forms the principal quarry of wildfowlers in the Northern Hemisphere. Most ducks eat small crustaceans or molluscs and vegetable material such as seeds, while a small number of species eat fish. Many species obtain their food by diving, and have streamlined bodies with webbed feet at the back of the body providing efficient thrust. Ducks fly fast and are capable of long-distance migration. Some of the Tufted Ducks *(Aythya fuligula)* which winter in Britain, breed in Siberia.

**Falcons**
The term 'raptor' includes the eagles, hawks, falcons and vultures, which together form the order Falconiformes. The Peregrine Falcon *(Falco peregrinus)* is a species which has captured the imagination of many. It flies very fast and can attack and kill in flight pigeons, ducks and other species as large as, or larger than, itself with one blow at the end of a stoop. The stoop is a very rapid descent, with partly closed wings, during which speeds of 130 km/h (81 mph) have been recorded using a miniature air-speed recorder. It seems likely that even greater speeds are sometimes achieved. The Peregrine Falcon is the most widespread species in the world. It breeds in arctic and tropical climates and might be seen in almost any part of the world.

**Pheasants**
The order Galliformes includes the most important of the birds which are used for food by man. The turkeys form one family, guinea fowl another, and grouse a third, but the largest is the pheasant family which includes the Domestic Fowl *(Gallus gallus)* as well as 164 other species of pheasant, partridge, quail, and peafowl. The bills of these birds, which are short and

Above
The Water Rail *(Rallus aquaticus)* is a secretive bird which walks around at the base of reeds. Some rails which live on islands have become flightless, a recently publicized species being the Takahe *(Notornis mantelli)* from New Zealand which was thought to be extinct until rediscovered a few years ago in South Island. Many of the most secretive rails are weak fliers, but a few fly well and migrate over long distances.

Right
Sandgrouse are related to pigeons rather than to game birds. The Pin-tailed Sandgrouse *(Pterocles alchata)* lives in arid areas of Africa and is a regular visitor to water-holes where, like pigeons, they drink without raising the head.

thick, are used to dig the superficial layers of the ground for vegetable food such as seeds, roots and bulbs, as well as worms and insect larvae. The legs and feet are strongly built for scratching the ground and, in many species, for fast running. Flight is usually limited to short distances, although the small quail are capable of migrating. Also included in this order is the South American Hoatzin *(Opisthocomus hoatzin)* whose young have two well-developed claws on each wing.

**Rails**
The members of this family have medium-length strong legs with large toes and fairly thickset bodies, for most of them spend their time creeping about in vegetation. There are 132 species whose English names often include the words rail, crake or gallinule. They live in marshy places or in damp places on land. The Moorhen or Common Gallinule *(Galinula chloropus)* occurs in most parts of the world except Australia, where its niche is occupied by the Dusky Moorhen *(Gallinula tenebrosa)*. Like most rails, it walks and swims with a characteristic bobbing motion.

**Sandpipers**
The Scolopacidae is the largest family of waders or shore birds with seventy species, and includes snipe, curlews, godwits and shanks, as well as sandpipers. They have long bills and legs, slim bodies, and well-developed wings for fast flight. Most are social and are vocal when feeding on mud or sand flats. The loud piping calls of Redshank *(Tringa totanus)* or the Lesser Yellowlegs *(Totanus flavipes)* are well known to European and North American ornithologists. The variety of bill and leg lengths of birds in this family is described in the next chapter in relation to the depths at which their prey live in the mud or sand. Related families, include the plovers and stone curlews, which are more terrestrial, and the oystercatchers, avocets and phalaropes.

**Gulls**
The characteristic sound of the coast in north-west Europe is the call of the Herring Gull *(Larus argentatus)*, and its niche is occupied in other parts of the world by a a variety of other gulls. Gulls are very adaptable birds. They will eat a wide range of different sorts of food and they are quick to exploit a new food source. Various species, including the Herring Gull, have become more terrestrial in recent years as man has started to produce larger and larger concentrations of refuse. The gulls compete with members of the crow family, rats and, sometimes, vultures for edible material deposited on refuse tips. At the same time, gulls have started to breed on man's buildings, in addition to their previous breeding cliffs, and the total populations of gulls have dramatically increased in many parts of the world. The terns in the same family and the related families of skuas and auks have benefited much less from man's activities, and some species have declined dramatically.

Below
Bustards have strong legs and are capable of running swiftly. The Great Bustard *(Otis tarda)* is a European and Asiatic species which is one of the heaviest flying birds, reaching as much as 16 kg (35 lb). The male displays by lowering its head and bringing its wings and tail forwards.

Bottom
The Sandhill Crane *(Grus canadensis)* breeds in the northern parts of the United States and in Canada. They migrate south to southern USA and Mexico for the winter. Cranes are placed in the same order as rails and bustards.

## Pigeons

The pigeons or doves form a large family of 289 species, whose only close relatives are the sandgrouse. Many of the species are very successful today and the Passenger Pigeon *(Ectopistes migratorius),* which became extinct in North America early this century due to man's depredations, may have been the commonest bird in the world 200 years ago. The pigeons and doves eat seeds, fruit, buds and other, principally vegetable, food. There are species which are common enough to be occasional or regular pests in most countries. The important pest in northern Europe is the Wood Pigeon *(Columba palumbus),* a very common species which is widespread in cultivated land and also occurs in towns where it mingles with the semi-domesticated town pigeons, themselves descended from the Rock Dove *(Columba livia).* A dramatic example of an expansion in range which has occurred without any deliberate introduction by man, is shown by the Collared Dove *(Streptopelia decaocto)* which first appeared in western Europe at the beginning of this century, but which has now spread to most parts of the Continent. It spread from Asia Minor and first appeared in Belgrade in 1912, in Germany in 1946, in Denmark and the Netherlands in 1948, in France and Poland in 1950, Moldavia in the USSR in 1952, Switzerland and England in 1955 and Scotland in 1958. Collared Doves are now very common breeding species in most parts of Britain, and it would be interesting to see how successful the species would be if it reached North America, where it would compete with the widespread Mourning Dove *(Zenaida macroura).*

Above
The African Grey Parrot *(Psittacus erithacus)* has been kept as a cage bird for centuries. The voices of parrots are usually raucous and the birds are gregarious so that flocks of parrots are very obvious. Their popularity as pets is due in part to their bright plumage, but also to the great ability of several species to mimic a variety of sounds including the human voice.

Bottom right
An intriguing member of the cuckoo family is the Greater Roadrunner *(Geococcyx californianus)* which lives in the south of the USA and in Mexico. This is quite a large bird, 55 cm (22 in) long with a long tail and strong legs. Roadrunners do run a great deal and prey on ground-living animals including snakes. Their movements when attacking a snake are rapid enough to allow them to deal with poisonous species such as rattlesnakes.

Right
The Cuckoo *(Cuculus canorus)* resembles members of the hawk family. Females about to lay eggs may use this resemblance to frighten hosts off their nests. The young Cuckoo is larger than the nestlings of the host species and ousts them from the nest.

### Parrots

The 315 species of parrots live in places with tropical and subtropical climates, especially in the Southern Hemisphere. They are immediately recognizable because the bill is always short, stout and strongly hooked with a bulging fleshy portion (the cere) at the base of the upper mandible. This mandible is articulated with the skull and is therefore movable and not fixed as in other birds. The food is principally fruit and nuts. Most parrots can fly fast for short distances, although they seldom fly far.

### Cuckoos

Although represented by only one species in northern Europe, the cuckoos are a large family of 127 species which has a worldwide distribution. The parasitic breeding habit, which is well known for the Cuckoo *(Cuculus canorus)* in Europe, is shown by forty-seven species of cuckoo as well as by a variety of other birds. Each female Cuckoo lays her eggs in the nests of one host species only, and her eggs normally resemble those of the host. Thus, Cuckoos are members of 'clans', and those females hatched and reared by, for instance, a Reed Warbler *(Acrocephalus scirpaceus)* will return to lay their own eggs in a Reed Warbler's nest. In an extensive and uniform habitat this adaptation is beneficial to the Cuckoo, but in a situation where many hosts live, Cuckoos from different 'clans' may hybridize and produce eggs which are conspicuously different from those of the host. Another intriguing member of the cuckoo family is the Greater Roadrunner *(Geococcyx californianus)* which lives in the south of the USA and Mexico. This is quite a large bird, 55 cm (22 in) long with a long tail and strong legs. Roadrunners do run a great deal and prey on ground-living animals including snakes. Their movements when attacking a snake are rapid enough to allow them to deal with poisonous species such as the rattlesnake.

## Owls

Most of the 130 species of owls are placed in one family, but the barn owls are a separate family. Their way of life is most similar to that of the birds of prey, Falconiformes, although the two groups are not closely related. Like birds of prey, owls have a hooked bill, large talons and forward-facing eyes. Most hunt at night and are extremely well adapted for nocturnal life. They range in size from the sparrow-sized Elf Owl *(Micrathene whitneyi)* of North America, to the eagle owls *(Bubo)*, from various parts of the world, which are as large as many eagles. All eat animal food, especially those mammals which are also nocturnal. Two genera of owls, *Scotopelia* in Africa and *Ketupa* in Asia, eat fish which they catch in the water rather as the Osprey *(Pandion haliaetus)* does. Most breed in old nests of crows and birds of prey, and it is during the breeding season that the hooting calls of the many species are most frequently heard. Many owls range over large parts of the world and the Barn Owl *(Tyto alba)* is second only to the Peregrine Falcon *(Falco peregrinus)* in the extent of its range.

The fringe of barbs at the front of the feathers of this Tawny Owl's *(Strix aluco)* wing prevents the production of the whistling noise which the wings of birds usually make. This allows owls to fly silently to their prey which they are able to detect at night by means of their acute sight and hearing.

## Nightjars

Most species of nightjars, and their close relatives the potoos, frogmouths, and the Oilbird *(Steatornis caripensis)*, are nocturnal. They have large eyes and barred and mottled plumage which camouflages them very effectively during the daytime. Most feed by catching insects in flight and, since many of these insects are quite large moths, they have extremely wide gapes which are usually fringed with bristles to guide in the insects. The Common Nighthawk *(Chordeiles minor)* of North America feeds at dusk and is agile in flight and falcon-like in silhouette but flies much more slowly than a falcon. The noises made by nightjars in the breeding season are well known in many parts of the world. The European Nightjar *(Caprimulgus europaeus)* makes a prolonged fluctuating churring sound while the Whip Poor-will *(Caprimulgus vociferus)* and the Poor-will *(Phalaenoptilus nuttallii)* of North America call their names throughout the night. It has recently been discovered that the Poor-will can hibernate in rock crevices during the winter. The Oilbird of northern South America and Trinidad is remarkable in that it is the only nocturnal fruit-eater. It prefers to feed on the fruits of the oil-palm and it navigates at night by means of a form of echo-location not unlike that of bats, except that the clicking noises which it uses are of a much lower frequency than are the ultrasonic calls used by bats.

## Swifts

The wings of swifts are long and narrow and their wing muscles are very well developed so that they can fly very fast for long periods. They glide for some of the time but their manoeuvres for catching their insect prey depend principally upon flapping flight. Their tails are reduced so the wings must be used for steering. The extreme efficiency of their flight allows them to spend the night on the wing, so many species may never land except for breeding. Even copulation and the gathering of material for the nest can be accomplished on the wing. A consequence of this aerial life is that the legs and feet, which are seldom used, are reduced in size so that they can do no more than cling on and shuffle along the

The nests of this Indian Swiftlet *(Collocalia fucifaga)* and of related species are formed principally of an edible cement. It is this which is used to make birds' nest soup.

ground with them. Swifts can take off from flat ground but not from low vegetation, so they normally land only on cliffs. It seems likely that a swift is the fastest moving animal; the Spine-tailed Swift *(Chaetura caudacuta)* of the USSR has been timed at 170 km/h (106 mph) in level flight. Like the Oilbird previously mentioned, some swifts live in caves and use echo-location. An example is the Black-nest Swiftlet *(Collocalia maxima)* whose most well-known nesting colony is the Niah cave of Sarawak where about two million birds nest.

### Hummingbirds

By virtue of their small size, fast flight, ability to hover, iridescent coloration, and association with flowers, the hummingbirds stand out as one of the most remarkable families of birds. All 320 species live in North or South America and many of them are little larger than the biggest insects. The Bee Hummingbird *(Mellisuga helenae)* of Cuba, and several other species are less than 7 cm (2·8 in) in length including the bill and tail. The wings, like those of swifts, are made up of primary feathers only and are rather narrow and pointed. As a consequence of muscular adaptation, and the great flexibility of the shoulder joint, hummingbirds can hover and can flap their wings very rapidly. The fastest recorded wing beat of any bird is 90 beats per second by the South American species *Heliactin cornuta*. Such fast rates of wing beating result in the

Left
The Collared Trogon *(Trogon collaris)* is much more frequently heard than seen. Its call is characteristic of many of the forests of Central and South America.

Right
The Kookaburra *(Dacelo novaeguineae)* of Australia lacks the brilliant plumage of most members of the family but its laughing call is very striking, especially when several individuals call in chorus. This individual has caught a lizard although the birds also eat many large insects.

production of a humming noise. The food of hummingbirds is nectar and small insects, and the tongue is adapted to suck nectar, for it is long and has a groove down the centre. The bird hovers in front of the flower and sucks from each nectary in turn before moving rapidly to the next flower. Since hummingbirds are small and can be fed nectar, their energy utilization can be studied more precisely than can that of other birds. Measurements of oxygen uptake at rest during day and night and during hovering flight, together with estimates of the time spent by the average bird in each of these activities, showed that the Anna Hummingbird *(Calypte anna)* from California needed to consume the nectar from about a thousand *Fuchsia* blossoms in order to provide for its energy needs. Hummingbirds in mountainous regions save energy at night by assuming a torpid state in which their temperature drops.

**Trogons**
These brightly coloured birds from Africa, and Central and South America are all the size of a large thrush or small pigeon and feed on fruit or insects in forests. The species with the most extravagant plumage is the Quetzal *(Pharomacus mocino)* of Central America which was the sacred bird of the Mayas and Aztecs. Its upper tail covert feathers may be almost 1 m (39 in) long and its plumage is a brilliant mixture of green, red and white.

**Kingfishers**
Another order of brilliantly coloured birds includes the families of todies, motmots, bee-eaters, rollers, hoopoes, hornbills and kingfishers. There are over eighty species of kingfishers ranging in length from 10–45 cm (nearly 4–18 in). The European and North American species eat fish which they catch by diving from a perch, for instance the Kingfisher *(Alcedo atthis)* of Europe and Asia, or by hovering and then diving from the air, as in the Belted Kingfisher *(Megaceryle alcyon)* of North America. More than half of all kingfisher species are entirely terrestrial and feed by catching insects on the wing or by flying down from a perch to catch large insects.

**Honeyguides**
These curious birds are included in the same order as the woodpeckers, barbets, jacamars, puffbirds and toucans. They are dull in colour and feed on insects and their waxy secretions. The members of the African genus *Prodotiscus* feed on scale insects, but the name of the birds derives from the habits of the Black-throated Honeyguide *(Indicator indicator)* which draws the attention of honey-badgers and man to bees' nests and then consumes wax and larvae when the nest has been destroyed. All honeyguides lay their eggs in the nests of other birds.

**Woodpeckers**
The feeding adaptations of members of this family of 230 species which are distributed throughout the world, are described in the next chapter. The whole of their anatomy is adapted for an arboreal life during which most time is spent perched on vertical

tree trunks. The tree trunk is used for nest excavation and, in some species, to make loud drumming noises which replace song as a means of proclaiming territorial ownership. Their ability to feed wherever there are trees allows them to live in cold mountain woodlands as well as in temperate and tropical forests. The largest species, such as the European Black Woodpecker *(Dryocopus martius)* and the Pileated Woodpecker *(Dryocopus pileatus)* of North America, are crow sized.

**Ovenbirds**

The large South American family of ovenbirds, spinetails, treerunners, leaf-scrapers, etc., is remarkable both for the diversity of habitats which its members utilize, and for the calls and nests of many members. No ornithologist in South America can fail to be aware of the calls of spinetails and various other species in this family, but due to their secretive habits and inconspicuous plumage the originator of the sound is seldom noticed. The 221 species of the Furnariidae are all less than 25 cm (10 in) in length but they include warbler-like species which creep among leaves seeking insects; dipper-like species which feed in or near streams or even on floating masses of seaweed; ground-running species; nuthatch-like species; and some which resemble the related family of woodcreepers in their habits. The name of the family is due to their nesting habits, especially that of the Ovenbird *(Furnarius leucopus)* whose large mud-oven nests are a characteristic sight in open agricultural areas of several South American countries. Many other members of the family build large nests. The largest is that of the White-throated Cachalote *(Pseudoseisura gutturalis)* which is a barrel-sized structure with a domed roof strong enough to support the weight of a man.

**Antbirds**

This family is also principally South American and most of those which have English names include the word 'ant' in their name. One bird which has achieved some fame is the Cock-of-the-Woods or Black-faced Ant-thrush *(Formicarius analis)* which is quite common in Trinidad and northern South America. It is frequently found in association with army ants in forests and feeds principally on insects and other small animals which are disturbed by the ant column. Some of the antshrikes such as the Black-crested Antshrike *(Sakesphorus canadensis)* feed like a small shrike and the males and females duet as do some African shrikes.

Above
The large South American family of antbirds includes over 220 species which are vocal and which often live in dense vegetation. The songs, like that of this Warbling Antbird *(Hypocnemis cantator)* may be familiar sounds but the birds are often difficult to see.

Right
The male Wire-tailed Manakin *(Teleonema filicauda)* has modified wire-like tail feathers which are used in display. Most members of this South American family gather at leks where the males display to one another prior to pairing.

**Cotingas**

The calls of bellbirds *(Procnias* species) and other Cotingidae are among the most remarkable to be heard in tropical American forests. The plumage of many members of this family is also bizarre, for in addition to striking patches of bright colour, males have crests, tufts of protruding feathers, or coloured wattles which are accentuated during the displays. The Cock-of-the-Rock *(Rupicola rupicola)* is completely orange in colour with a helmet-like crest extending over the bill and prolonged secondary feathers on the wings. Its nest is made of mud fixed to a rock face and the birds display socially and often nest close together. Another species with a crest extending over the top of the bill is the Umbrella-bird *(Cephalopterus ornatus)* which also has an enormous feathered wattle hanging from its chin so that it extends well below the perching bird.

**Tyrant Flycatchers**

Many of the species in this North and South American family feed by catching flying insects in very much the same way as do Old World flycatchers (Muscicapidae). Some of the larger members feed like shrikes and other smaller species feed on insects found while creeping among leaves, as do warblers. With a total of 365 species, this is one of

the larger families of birds and, due to the habit of perching in conspicuous places shown by many members of the family, they are among the most obvious birds throughout much of North and South America. Their songs are simple but are often easy for us to imitate so that several have names which are onomatopoeic. The peewees *(Contopus)* of North America call their name, as does the Kiskadee *(Pitangus sulphuratus)*, a wide-ranging species whose name is derived from the French *qu'est-ce qu'il dit*. The 'quick three beers' call of the Olive-sided Flycatcher *(Nuttallornis borealis)* is commonly heard in western North America.

**Swallows**

The seventy-four species of swallows and martins are the nearest equivalent to the swifts amongst the perching birds (Passeriformes). They are slower and more erratic in flight than swifts and are not able to fly continuously. Their habits of building their nests in close proximity to man's dwellings and arriving in temperate breeding areas in early spring have endeared them to many people not otherwise interested in ornithology. The Swallow *(Hirundo rustica)* which is called the Barn Swallow in North America, and the Purple Martin *(Progne subis)*, for which Americans build 'martin houses', readily nest on ledges in buildings. The House Martin *(Delichon urbica)* of Europe and the Cliff Swallow *(Petrochelidon pyrrhonota)* build mud nests affixed to the eaves of houses. Several species migrate across the equatorial regions twice in each year.

**Crows**

Most of the 102 members of the Corvidae are omnivorous, fairly large birds. Their adaptability and the achievements of captive individuals which have been presented with tasks and puzzles with food as a reward have resulted in a reputation for great mental ability. The fact that they often hold food with the foot while eating it and that they can manipulate well with their feet, even to the extent of jays pulling up food suspended on the end of a piece of string, has further enhanced man's opinion of their ability. It might be said that they are the most human of birds and this undoubtedly results in our holding a high opinion of them. Several species are social throughout their lives and show complex social organization which is by no means completely understood. The ravens are the largest passerine birds and, like many other crows, most are predominantly black in colour. Many of the species of jay are brightly coloured and these are the only crows which are found in South America.

**Birds of paradise**

New Guinea is the home of many of the forty-two species of birds of paradise but others occur in islands nearby and in Australia. They are arboreal birds which feed on fruit and on small animals living in the trees. The beautiful plumes, ruffs and patches of coloured feathers of the males have evolved in association with elaborate displays which are often carried out at communal display grounds. A few species have inconspicuous or completely black coloration and look similar to crows.

**Tits**

The tits and chickadees, like the crows, are great manipulators and problem-solvers, but they are all small birds and are very active and acrobatic in their searches for food. They are largely insectivorous but some species also eat nuts and they occur in all the continents except South America and Australia. The best known North American species is the Black-capped Chickadee *(Parus atricapillus)* which is very similar to the European and Asiatic Willow Tit *(Parus montanus)*. Several tit species produce domed lined nests with side entrances.

Left
The handsome Fork-tailed Flycatcher *(Muscivora tyrannus)* is a member of the large American family of tyrant flycatchers. It feeds like an Old World flycatcher by flying out from a perch to catch flying insects.

Above
The larks are known to the general public for their songs. The Skylark *(Alauda arvensis)* and the Woodlark *(Lullula arborea)* of northern Europe, and the Bifasciated Lark *(Alaemon alaudipes)* of north Africa and Asia sing varied and beautiful songs. Most are ground dwellers and several species occur in desert conditions. Wherever they live, their plumage resembles the colour of the soil.

Right
Magpie-larks *(Grallina cyanoleuca)* are common birds near water in Australia. They build large nests of mud with some grass embedded in it which may weigh as much as 1 kg (2·2 lb).

## Wrens

The common Wren *(Troglodytes troglodytes)* of Europe, which is called the Winter Wren in North America and which also occurs in Asia, is the only world-ranging species of a family of sixty species which is otherwise confined to America. This species is one of the smallest in the family and is insectivorous and yet, as its American name implies, it can survive the winter in temperate regions. In northern Europe, where most Wrens are non-migratory, hard winters do reduce their populations. Most wrens have loud songs and many build nests in which to roost as well as nests for egg-laying.

## Thrushes

The 300 species of thrushes are now usually grouped as a subfamily of the very large Muscicapidae family which also includes babblers, warblers and flycatchers but some authorities consider each of these groups a separate family. Many of the species of thrush look like the common garden thrushes of the Northern Hemisphere such as the Song Thrush *(Turdus philomelos)* of Europe and north-western Asia. Medium-sized worm- or snail-eating species can be successful in the tropics as well as in temperate regions. Most of them sing attractive songs and also eat fruit or insects. The other group of thrushes, the chats, have similar bodies and more slender legs. Some of these birds are called robins, wheatears or bluebirds and one genus, the nightingales *(Luscinia),* includes some of the finest songsters.

## Babblers

The babblers are a heterogeneous group of fairly small insectivorous birds which are found throughout tropical and subtropical parts of the world with the exception of the Americas, where the Wren Tit *(Chamaea fasciata)* from California is the only representative. The majority of the 258 species are brownish birds which call frequently as they move around fairly dense vegetation in groups. Some species, such as the laughing thrushes of eastern Asia, are brightly coloured and jay-like while others somewhat resemble tits, warblers, larks, wrens or even game birds. Species in the genus *Turdoides* of western Asia have been studied because of their complex social behaviour. Like some other babblers, these birds live socially throughout their lives. In the non-breeding season they move around in parties of six to twenty birds. They have an extensive vocabulary of calls which are of particular use in a species such as this which feeds on the ground in shrubby areas and in tall grass. They build nests within the feeding territory of the group and the young birds grow up with other members of the group around them as well as their parents. The social structure of the group is a hierarchy which is maintained by individuals offering gifts of food to those below them in rank order. The social behaviour of these birds is perhaps more sophisticated than that of any other animals except for some monkeys and apes.

## Warblers

The Old World warblers are generally considered as a subfamily, the Sylviinae, and the Australian warblers another subfamily, the Malurinae, of the large family Muscicapidae. The Sylviinae are represented by one boreal species of warbler and by kinglets in North America and are not closely related to the American warblers in the family Parulidae. They are slim birds with small thin bills, for their diet is almost exclusively insectivorous and they feed by creeping or flitting around in vegetation taking small insects from the surfaces of leaves. Many breed in temperate regions and therefore need to migrate to warmer countries in the non-breeding season. There are 320 species and they inhabit almost all types of vegetation. The majority are inconspicuous in coloration and in habit but the males sing distinctive songs. Some pairs of species such as the Chiffchaff *(Phylloscopus collybita)* and the Willow Warbler *(Phylloscopus trochilus)* of northern Europe are extremely similar in appearance but are readily distinguished by their songs.

## Flycatchers

The subfamily Muscicapinae is in the same family as the warblers, babblers and thrushes and includes 286 species. These are distributed all over the world except for the Americas where their place is taken

One of the largest species of wren, the Cactus Wren *(Campylorhynchus brunneicapillus)* of southern North America, builds large flask-shaped nests in thorny bushes or in cactus plants. Each individual builds its own nest soon after beginning independent life, and this serves as its home and protects it against the cold at night.

by the tyrant flycatchers already mentioned. The majority of flycatchers are small active birds, like the European and Asiatic Spotted Flycatcher *(Muscicapa striata)*, which catch their insect prey in flight. Like the Tyrannidae, however, the family has diversified in some places so that some members have quite different ways of life. In New Guinea there are small warbler-like species, chat-like species, and shrike-like species.

**Wagtails**

Wagtails and pipits are to be found in almost all parts of the world from the tropics to arctic regions and high mountain tops. Most of the forty-eight members of the family are either wagtails *(Motacilla)* or pipits *(Anthus)*. They are small terrestrial birds which feed on insects and other small animals. Many species live in damp or wet places. They run fast on the ground for short distances, often wagging the long tail up and down as they walk, but they are also strong fliers. Most species feed in loosely associated flocks for at least part of the time but their distribution depends upon the distribution of the food. Wagtails usually roost communally in reed beds or, in association with man, on factory roofs, in trees in cities, or at sewage works. While wagtails are often conspicuously coloured, most pipits are brown and streaked or spotted. They are characteristic birds of open grassland and moorland. Several species live in mountains up to 3,000–4,000 m (10,000–13,000 ft) altitude.

Right
The White-bearded Honeyeater *(Meliornis novaehollandae)* can often be observed feeding on nectar, pollen and insects at *Banksia* flowers in Australia.

**Shrikes**

The methods of feeding utilized by many shrikes (Laniidae), are further discussed in the next chapter. Most of the seventy-four species behave rather as small birds of prey, for they are active hunters of small animals. Many are brightly coloured and most have harsh calls. The African bush shrikes *(Laniarius)*, however, have attractive whistling songs which have recently been the subject of much study. It has been discovered that the male and female sing antiphonally. That is to say, they sing a duet in which they alternate notes or phrases of song. The resultant song was long thought to be made by the male alone, for the birds usually sing from areas of dense undergrowth. Another noteworthy group of African shrikes is the helmet-shrikes (e.g. *Prionops)*, which are very social birds and carry out curious display flights while they are among their group.

Left
The Great Grey Shrike *(Lanius excubitor)* is one of the larger members of its family and is capable of catching and eating small birds and mammals as well as insects. The bill is hooked and the birds frequently sit on the tops of bushes watching out for prey.

### Starlings

The immense success of the Starling *(Sturnus vulgaris)* in recent years has been possible due to man's activities, for they have increased in numbers in areas of intensive agriculture within their previous range in Europe and Asia and also in North America where they were originally introduced by man. The spread of the Starling across North America is well documented. The real explosion of numbers has occurred wherever Starlings have found areas of intensive, preferably mixed farming. The Starling is probably the commonest species in North America now. Their success must be due especially to their social behaviour as well as to their adaptability to the sort of vegetation patterns created by man. The other 109 species in the family are found in Africa and Asia and include several other highly successful species. The basic coloration of most starlings is black but many have iridescent plumage which gives them brilliant, metallic colours. One of the most well-known members of the family is the Hill Mynah *(Gracula religiosa)* of south-east Asia which is popular as a cage bird because of its remarkable ability to mimic the human voice (see chapter on conservation and economics).

### Honeyeaters

This, largely Australian, family of 167 species includes members with a wide variety of different ways of life. They range in size and body form from small species which are similar to kinglets or warblers, to larger birds which are like bee-eaters or magpies. Some of those intermediate in size resemble sunbirds, honeycreepers, or hummingbirds, while others look and act like flycatchers or orioles. They are confined to places where there are trees and eat nectar, pollen, insects, or fruit. Most are dull-coloured and gregarious.

The Red-whiskered Bulbul *(Pycnonotus jocosus)* is a common garden bird in India. There are 120 species of bulbuls in Asia and Africa and all are drab, noisy birds about the size of a House Sparrow or Starling. Most species eat fruit or other vegetable material but a few are insectivorous.

Right
This Lesser Double-collared Sunbird *(Cinnyris chalybens)* lives in southern Africa. Most of the 106 species of sunbird live in Africa or tropical Asia and are small and brightly coloured. They feed on insects or spiders and on nectar which they usually obtain by perching on flowers and probing with their long bills, for they lack the hovering ability of hummingbirds.

Below
Most white-eyes are small yellow-green birds with a ring of minute silky white feathers around the eye, like this Kapi White-eye *(Zosterops capensis)*. They are usually seen in flocks moving through bushy areas calling to one another and they eat insects, nectar and fruit. The eighty-five species, which are found in Africa, Asia, and Australasia, are very similar to one another and the task of classifying the members of the family has proved especially difficult.

## American warblers

The family Parulidae includes 113 species whose ways of life are similar to those of the Sylviinae warblers in Europe and Asia. They are all small birds, the majority being no larger than the Willow Warbler *(Philoscopus trochilus)* of northern Europe, and none being larger than a House Sparrow. They live and feed in trees or other dense vegetation by creeping around finding insects. The migration of these birds from temperate breeding grounds to warmer wintering areas often leads to spectacular 'falls' of migrants, especially in coastal regions, and the sudden advent of parties of warblers, including up to a dozen species, is one of the exciting ornithological events of each year for many North American ornithologists.

## Icterids

The American orioles, blackbirds, cowbirds and meadowlarks are in one diverse family and are often grouped together under the name 'icterid' derived from the family name Icteridae. They are medium to large birds which are similar in their range of ways of life to Old World starlings, orioles, crows and larks. Many of the eighty-seven species are tropical, but the Rusty Blackbird *(Euphagus carolinus)* lives in arctic tundra regions and many of the other northern species are also migrants. The grackles are omnivorous and often crow-like and will eat various small animals as well as fruit and seeds. Most of the orioles are vegetarian and several tropical species move around in flocks. The nests of orioles are frequently built in groups and those of the oropendolas are beautifully woven suspended nests up to 2 m (6·5 ft) long with apertures on the side. The meadowlarks *(Sturnella)* and the Bobolink *(Dolichonyx oryzivorous)*, in contrast, nest on the ground. The genus of cowbirds *(Molothrus)* includes species which range from *Molothrus badius* of South America which will sometimes build its own nest to a variety of other species which are brood parasites and lay their eggs in the nests of other birds.

## Tanagers

The subfamily Thraupinae which includes 191 species of tanager from the Americas, together with the buntings, American sparrows, cardinals and honeycreepers form the large family Emberizidae. The ornithologist visiting tropical Central or South America is immediately impressed by these common, brilliantly coloured birds. The

smallest are only 10 cm (4 in) in length while the largest are twice as long, about the size of a Starling. The majority are fruit eaters but some eat insects. A few species migrate north into temperate regions to breed and the male Summer Tanager *(Piranga rubra)* is yellow in the winter but bright red in its breeding grounds.

**Buntings**

The remainder of the family Emberizidae includes birds called by a wide variety of names but both the subfamily Emberizini, which includes 197 species of buntings and American sparrows; and the subfamily Cardinalinae, with 110 species including other birds called buntings as well as cardinals, have stout bills which are adapted for seed eating. They are about the size of a House Sparrow, or a little larger, and some, such as the Painted Bunting *(Passerina ciris)* and the Cardinal *(Pyrrhuloxia cardinalis)* are brightly coloured. Some North American sparrows are of particular interest because of their genetic variation and others because of their songs. The Song Sparrow *(Melospiza melodia)* is one of the most widespread species in Canada, the USA, and Mexico. The northern forms are considerably larger than the southern forms and there is a continuous series of intermediates. The darkness of the coloration and the amount of steaking also vary considerably from one subspecies to another. The songs of this species vary from one region to another but this type of variation has been most studied in the White-crowned Sparrow *(Zonotrichia leucophrys)*. This species has been recorded on Fletcher Island which is only 250 km (155 ml) from the North Pole and the song of the species varies from its northern breeding sites to the southern part of its breeding range in California to such an extent that a specialist can distinguish the songs of birds from areas as little as 100 km (62 ml) apart.

One of the commonest American warblers is the Yellow Warbler *(Dendroica petechia)* which, due to its all-yellow plumage and characteristic song, is easy to identify. It is remarkable among small insectivorous birds in that its breeding range is very large, extending from Alaska and Labrador to mangroves around the south and west of the Caribbean and to Ecuador and the Galápagos Islands.

Right
The Zebra Finch *(Poephila castonotis)* is an Australian member of the waxbill family. Most of the 107 species are small and eat seeds, but some are insect-eaters. The bright plumage of the males is often shown off in rapid-moving displays, during which the whistling songs may also be sung. Some of these songs are very complex and when tape recordings of them have been slowed down, several melodic lines have been found to be sung simultaneously.

Below
The House Sparrow *(Passer domesticus)* is successful wherever man lives in any numbers and has spread around much of the world. True sparrows and weavers form one family which includes 132 species and enormous numbers of individuals.

**Finches**
The Fringillidae are seed eaters and are found throughout the world except for Australia. Some, such as the Chaffinch *(Fringilla coelebs)* of Europe, are very common in a wide variety of temperate woodlands and do well in cultivated areas. Others, such as the Pine Grossbeak *(Pinicola enucleator)* of northern Asia, Europe and America, are confined to one sort of food and their distribution is correspondingly limited. The family is commonest in high altitude or temperate forests, but a variety of species is able to exploit tropical habitats. The song is frequently quite complex and the Canary *(Serinus canaria)* is a popular cage bird.

**Sparrows**
True sparrows and weavers, family Ploceidae, nest in large colonies and spend all or most of their lives in flocks. The extreme example, in terms of numbers, is the Quelea or Black-faced Dioch *(Quelea quelea)* of Africa. Several other weavers and sparrows are very common in agricultural areas where they feed on seeds and may be serious pests. The communal nests of some weavers are enormous. One colonial dwelling of the Social Weaver *(Philetarius socius)* was 10 m (33 ft) long, 7 m (23 ft) wide, and 1·5 m (5 ft) high. The House Sparrow is also a sociable species but is most noteworthy for its ability to utilize man's buildings and waste food so that it is found in urban areas.

# How birds feed

This group of White-backed Vultures *(Gyps africanus)* are feeding on the carcass of a dead Lechwe, an African antelope. Vultures soaring over the plains watch one another and if one sees a dead animal and descends to it, others quickly converge on the point.

Food, once digested, provides energy which must be sufficient for all the activities of the animal. Energy is, however, consumed during the active processes of feeding and digesting food. Animals cannot afford to waste energy when looking for or consuming food, so feeding methods must be efficient. Since birds exploit a wide variety of different sorts of food their feeding methods and, as a consequence, their anatomy, have become adapted in many ways.

The first problem is to find the feeding place. Most birds live close to potential food sources, but when they exhaust the food supply in one place they need to find another. A flock of tits may have to move from copse to copse in their search for tiny insects on trees. Another reason why birds may not be able to stay at a feeding place is that it may sometimes be hazardous due to the presence of predators. This is a problem for almost all ground-feeding birds which cannot stay in the feeding area at night. Once a potentially profitable feeding area is found, the individual food items must be located and captured. For some species the food items are concealed or difficult to extract; for others the food is easy to detect but difficult to catch. At each stage effort is required, so special behavioural strategies have evolved. For many species the food finding or catching procedure is more efficient if carried out by a flock. The particular advantages of flocking behaviour are stressed in the next chapter.

If an animal can find sufficient food to last it twenty-four hours within a few minutes each day, as in the case of a falcon which makes a kill soon after starting to hunt, it does not need to hunt at all for for the rest of the day. A large bird of prey may be able to obtain enough food from the capture of one prey animal to last it for several days. Birds which feed on very small prey, or which are hunting unsuccessfully, may need to continue directing most of their energy to obtaining food for most of their waking hours. There is often a seasonal variation in the availability of food; thus insects are much harder to find in a prolonged winter or dry season. Many small birds which have to work hard to obtain enough food during the unfavourable season, work even harder in the favourable season because they breed at this time and they have to provide for their mates or for a nest full of voracious young. The animals are thus spending most of their time working at almost maximum capacity and such species tend to have short lifespans. Birds of a similar size which live longer, for instance petrels and many other seabirds, may not have to work hard for their food.

The general strategies adopted by birds hunting for food depend upon the distribution of the food sources. Most food is not evenly distributed; some is concentrated in dense but widely spaced patches. Bivalve molluscs such as mussels are a food source for oystercatchers *(Haematopus)* and eiders *(Somateria)*. They may occur scattered over rocky outcrops in the tidal zone and in shallow water, but in some localities they occur in very large aggregations. The birds can survive best if they concentrate their feeding efforts on these dense mussel beds, so it is essential for each individual bird to find an accessible mussel bed. Similarly, terns *(Sterna)* must find schools of small fish. Finches, buntings and sparrows must find areas where there are concentrations of seed-bearing plants. Fruit-eaters such as tanagers and toucans must search for trees which are in fruit. Birds feeding at a site where food is concentrated must decide on the best time to move on to another feeding site. This decision may come at a time when so much of the food has been eaten that it requires too much effort to find each new food item. It may, however, be forced upon the bird by nightfall, an incoming tide, the arrival of competitors or a predator. If competitors arrive each bird may have to expend energy defending any food item procured so its food-gathering efficiency may be best maintained by moving to another feeding site. In all situations where predation is a possibility, the bird has to balance advantages associated with continued feeding against the probability of being caught by a predator.

A bird which is within reach of food still has to decide on the strategy which will allow it to obtain the maximum amount of food with the least possible expenditure of energy. A long-legged fish-eater such as a heron may be most successful if it stands quite still and aims a thrust of its spear-like bill at fish which are within its reach. A wading bird walking on mud might obtain the maximum amount of food by probing with its bill in every little hole which is visible on the surface of the mud. A flamingo which sieves crustaceans from water with its curiously shaped bill, must catch most of its food without seeing it first. Most birds, like the wader and the heron, depend upon recognizing food items or hiding places of food items, before making a movement which might result in obtaining food. The ability of birds to recognize their food and to husband their energy by refraining from useless movements will improve with practice.

Young birds are much less efficient than adults and are consequently less likely to survive. The difference between young and old is most obvious in those species which need great skill to find or catch prey, for instance birds of prey. Much effort is required to chase active animals, so a large number of unsuccessful chases is very costly in terms of energy utilization. One consequence of feeding on food which is difficult to obtain is that the young are fed by the parents until they are much more developed than in those species which can obtain their food easily.

One intriguing example of a difficult feeding method is the oystercatcher which has already been mentioned. The European Oystercatcher *(Haematopus ostralegus)* eats bivalves, especially mussels. The mussel *Mytilus edulis* has a hard shell and a strong muscle holding the two valves of the shell together. Oystercatchers attack the mussel in one of two ways, each individual bird specializing in one of the methods. Some Oystercatchers are stabbers: they stab between the valves of the shell and cut through the muscle, thus allowing the opening of the shell. If the muscle is not hit with the first stab, it will close the valves of the shell together and may trap the bill of the bird. Other Oystercatchers are hammerers: they find the precise spot where the muscle is located and hammer a hole through the shell with the bill. If the hole is made in the wrong place the bird will have wasted much energy to no avail. Both of these

Left
The White Stork *(Ciconia alba)* uses its long strong bill to catch frogs. It also feeds on fish and snails in marshy areas, and on large insects in fields.

Below
Avocets *(Recurvirostra avosetta)* feed by swinging the head from side to side and sifting small crustaceans and other animals from the water with the upturned bill.

feeding techniques require very precise co-ordination of vision and movement and they are too difficult for young birds. The parents therefore continue to feed the young until they are full-sized adults. Oystercatchers cannot feed with enough efficiency to allow them to embark upon the demanding process of breeding until they are three years old.

Some species of birds are very specialized in their diets. The Snowy Owl *(Nyctea scandiaca)* and the Long-tailed Skua *(Stercorarius longicaudus)* feed almost exclusively on lemmings in their arctic breeding areas, and are inhibited from breeding if these small mammals are absent. Two other specialist feeders are the Snail Kite or Everglade Kite *(Rostrhamus sociabilis)* and the Limpkin *(Aramus guarauna)* of Florida and the northern parts of South America. Both feed on the freshwater snail *Pomacea*. At the other extreme are many species of crows and gulls which are very catholic in their feeding habits, taking various kinds of animal and vegetable material. Most species prefer one type of food but will readily change to another if the first becomes locally scarce. Sometimes the same family includes some species which concentrate on a single food source and others which do not. The diet of the Red Grouse *(Lagopus scoticus)*, which is confined to Britain, was found to be 74 per cent leaves, fruits and seeds of heather. The American Ruffed Grouse *(Bonasa umbellus)*, on the other hand, ate 374 kinds of plants and 131 kinds of animals and of the sixth most important foods of this species in northern Ohio, only two are found in the list of the top six foods in southern Ohio.

# Specializations for feeding

No living birds have teeth, so they must use other means to break up the food into small enough particles for digestion to occur. The beak is hard enough in most species for pieces to be pecked off soft food materials, such as fruit, and birds of prey use their hooked bills and talons to tear strips of meat off carcasses. Most birds, however, swallow considerably larger lumps of food than would a mammal of the same size. Further breakdown of food occurs in the muscular gizzard. Stones and other hard particles accumulate in the gizzard, which has a leathery lining, and lumps of food are broken down by squeezing and grinding movements brought about by the muscles in the gizzard wall. Further breakdown of food is effected by enzymes which are secreted by other parts of the gut. Materials like hair, bones, insect wing-cases, etc., which are resistant to digestion, are often aggregated into pellets and ejected from the mouth rather than passed into the intestine. Such pellets are also known to be produced by insect-eaters as well as by birds of prey and owls.

A further specialization of the gut is the crop, which is the sac into which food passes after being swallowed. The crop is highly developed in grain-eaters and other vegetarians, for its function is principally storage. Birds can digest animal food very rapidly, but they are much less well adapted to digest most vegetable material. They therefore need to consume a larger volume of plant food in order to gain sufficient nourishment from it. Plant food is often found near the ground in places which are

Left
The Willow Warbler *(Phylloscopus trochilus)* is typical of many small insectivorous birds in its bill and body shape. It feeds by creeping along small branches and among leaves, picking small insects off them.

Below
The Bald Eagle *(Haliaeetus leucocephalus)*, which is the national bird of the USA, feeds principally on fish which it catches by sweeping down from the air and grasping with its strong talons. The bill is heavy enough to pull pieces of flesh from a large fish which the bird holds down by the feet.

hazardous due to the presence of ground predators. Due to the presence of the crop, when suitable seeds, leaves, or other foods are found the bird can eat rapidly, fill its crop, and then retreat to a safer place for the prolonged digestive period. The anatomy of the bird is thus well adapted to fit in with its essential behaviour.

**Flesh-eaters**

The most obvious specializations for feeding shown by birds are those of general body anatomy. Birds which need to wade in water to find their food need long legs, long toes so that they do not sink in mud, and long necks so that they can reach the ground without bending their legs. Birds which chase fast-flying insects or other birds need to have wings and tails adapted for rapid flight and fast turning. Those which chase fish by swimming under water need a streamlined body and paddle-like feet or wings. An appropriately shaped beak is needed for the act of catching hold of the food item. All these anatomical characteristics are of use only if the bird also has an efficient behavioural strategy so that it can reach the prey efficiently. A variety of anatomical and behavioural feeding specializations are detailed below.

Birds of prey have hooked bills and talons on their feet so that they can grasp and rend their prey. Golden Eagles *(Aquila chrysaetos)* have talons and a bill large enough to deal with hares, their principal prey. The feet of snake-eating eagles such as the Short-toed Eagle *(Circaetus gallicus)* have short toes

carrying large bones and drop them on to rocks beneath. The bones smash and the vulture can descend and eat the bone marrow. A similar habit is shown by vultures which drop stones on Ostrich eggs so that they can eat them and by gulls which drop crabs or molluscs on to rocks so that the hard carapace or shell is broken.

**Insect-eaters**

The smallest falcons, hawks and owls eat insects, and they tend to resemble other large insectivorous birds in anatomy and behaviour. The hook on the upper mandible of the bill is less pronounced and the bill is slighter, but not necessarily shorter, than that of a flesh-eater. The talons are not large but they may be used to hold large insect prey while it is dismembered, for instance by tyrant flycatchers as well as by falcons and owls. Large insects are usually eaten at a perch near the place where they are caught, but the shrikes may carry the prey to a 'larder' in a thorn bush. The prey, whether insect, small mammal, or small bird, is impaled upon a thorn and then eaten immediately or kept until later. Shrikes and large flycatchers usually feed principally on large insects, such as grasshoppers, which they catch on the ground. Smaller flycatchers rest on perches like the larger species, but actively chase and catch flying insects. They usually return to the perch to remove the wings before eating the insect. Even smaller flying insects are eaten by swallows, martins and swifts. These birds remain on the wing for long periods and have small bills, but a large gape.

Among the most abundant insects in woodland are aphids and related insects which live on plant stems and suck their sap, and caterpillars and sawfly larvae which live on leaves. Both of these are eaten by small insectivorous birds such as warblers in the Old World and Parulidae in the Americas, and tits or chickadees. Tits will sometimes enlarge crevices in tree trunks in order to reach insects, but this habit is more suited to the nuthatches, which have strong sharp bills for making holes, and the Tree Creeper or Brown Creeper *(Certhia familiaris)* which has a long curved bill which can be inserted into crevices. The animals hidden deeper under

and are very strong for the prey must be grasped and killed very rapidly. Another snake-eater is the Secretary Bird *(Sagittarius serpentarius)* of central and southern Africa which has long legs and spends much of its time striding about in grassland. It kills snakes by stamping on them, so its toes and claws are short but hard. The feet of the fish-eating Osprey *(Pandion haliaetus)* and of the fishing owls *(Ketupa* species), have sharp curved claws and spiny tubercles under the toes so that their slippery prey can be grasped when they swoop down to a victim swimming near the surface.

An alternative to the eagles' means of obtaining animal flesh for food is to eat animals which are already dead. In some parts of the world there are many mammals too large for any bird to kill, so in order to exploit the food source the birds must depend upon mammalian carnivores to kill them first. The main groups of birds which feed on carrion are the two families which include vultures, one in the Old World and one in the Americas. Other carrion-eaters include crows, gulls and the Marabou Stork *(Leptoptilos crumeniferus)* of Africa. Some vultures have developed special behavioural techniques to deal with parts of the dead animal which would otherwise yield no food. The Lammergeier *(Gypaetus barbatus)* will fly up in the air

Left
Most bee-eaters catch bees and wasps in flight, but this Rainbow Bird *(Merops ornatus)* has caught a dragonfly.

Above
Moths are easily snapped up in the enormous gape of this Red-necked Nightjar *(Caprimulgus ruficollis)*. The sight of the gape may also be intimidating to interlopers at the nest site.

Right
The Marabou Stork *(Leptoptilos crumeniferus)* differs from others in its family in its diet. It is a carrion-eater and uses its bill to probe in carcasses of large animals and eat the entrails. Like many vultures, the head and neck are largely bare of feathers, for they would otherwise become matted with blood.

Below
The bill of the Wood Sandpiper *(Tringa glareola)* is of medium length and the bird feeds by walking on the mud probing in holes made by worms and crustaceans.

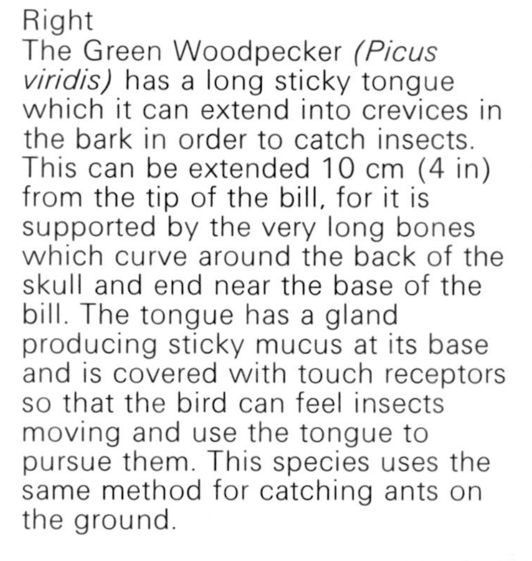

Right
The Green Woodpecker *(Picus viridis)* has a long sticky tongue which it can extend into crevices in the bark in order to catch insects. This can be extended 10 cm (4 in) from the tip of the bill, for it is supported by the very long bones which curve around the back of the skull and end near the base of the bill. The tongue has a gland producing sticky mucus at its base and is covered with touch receptors so that the bird can feel insects moving and use the tongue to pursue them. This species uses the same method for catching ants on the ground.

the bark of trees can be reached by woodpeckers which have chisel-like bills and strong neck muscles so that they can hammer holes in tree trunks. Hammering on a vertical surface is possible only if the bird can cling securely to the tree trunk. Woodpeckers have strong feet with four toes, two of which face forwards and two of which are backward facing, in addition to a stiff tail which acts as a prop. They reach the insect by means of their long sticky tongues. A remarkable similarity exists between the feeding of woodpeckers and that of the Galápagos Woodpecker Finch *(Camarhynchus pallidus)* which has no long beak and tongue, but which extracts insects from deep crevices in bark by means of a cactus spine. This is an example of a bird using a tool. Many other birds eat insects, spiders and other small animals on the ground and in low vegetation. Since insects are by far the commonest land animals it is not surprising that they are the principal food of the majority of birds. Even grain-eating birds like House Sparrows *(Passer domesticus)* and Grey Partridges *(Perdix perdix)* eat, or are fed, insect food while young.

**Worm- and snail-eaters**

The birds whose feeding is easiest to observe in gardens are those which feed on worms and snails. The Blackbird *(Turdus merula)* and Song Thrush *(Turdus philomelos)* of Europe, and the American Robin *(Turdus migratorius)* are sufficiently large and common to be obvious to the most casual observer. The Blackbird and the American Robin are very similar in every way except in plumage, and both hunt earthworms. They are probably able to detect the movements made by earthworms near the surface of the ground by means of vibration receptors in their legs. When the bird turns its head on one side it is probably focussing the most sensitive part of its eye on a worm hole. Once seen near the surface, the worm is grasped and pulled out of its hole. The Song Thrush is one of the four species of birds which can eat land snails. The snail is carried to a stone, known as the thrush's 'anvil', and its shell is smashed by twisting the neck and banging the snail on the anvil. If the shell is not held in the correct place, it will not break.

Molluscs are important food items for various marine birds. As previously mentioned, mussels are eaten by oystercatchers, eider ducks and gulls. Many waders, or shorebirds, feed on molluscs which often burrow beneath the surface of the sand or mud, so the bird must probe with its bill. The Shelduck *(Tadorna tadorna)* sieves surface mud and removes the tiny snail *Hydrobia*. These are sometimes very common and may be eaten by various waders such as the Dunlin *(Calidris alpina)*. The bivalve *Macoma* lives at 2–5 cm (0·8–2 in) beneath the surface of sand and is the favourite food of the Knot *(Calidris canutus)* which has a bill which is just long enough to reach the mollusc. The shorebird probes every hole on the beach and is able to extract the molluscs at a rate of about 750 per day. Among other very common small animals living on the shore are worms and crustaceans. The two species of turnstone of the genus *Arenaria* find small crustaceans by lifting stones and seaweed with their bills and catching the individual animals hidden beneath. Most curlews, godwits, shanks and sandpipers catch small crustaceans and worms by probing in mud.

**Fish-eaters**

The surface layers of the sea are inhabited by many small crustaceans, squids and fish which feed on planktonic animals or plants.

Right
Several members of the heron family, like this Louisiana Heron *(Hydranassa tricolor)*, catch fish by stabbing at them with their long sharp bills. The heron keeps its body still when stabbing so that the fish see a minimum of movement.

Below
The Red-breasted Merganser *(Mergus serrator)* has a saw-edged bill with a hooked tip, with which it catches the fish it chases underwater. In this respect it differs from most other ducks which feed on crustaceans, molluscs and vegetable material such as seeds.

These form the food of some of the most numerous birds in the world, the petrels and shearwaters. These birds spend most of their lives far from land. The albatrosses and large shearwaters, which are also oceanic, eat larger squid and fish which they catch on the water surface, often while still in flight. In coastal waters, many different sorts of birds feed on fish. The divers or loons, grebes, penguins, cormorants, some auks, and ducks all dive from the surface of the sea and chase fish under water. All are streamlined and have fairly long bills suitable for grasping fish. They have webbed feet and some have wings which are also adapted for swimming. The throat is usually distensible so that large fish can be swallowed.

An alternative method of catching fish is to dive from the air and catch fish swimming just below the surface. This method is used by gannets and boobies, the Brown Pelican *(Pelecanus occidentalis)*, and terns and kingfishers. The sight of groups of gannets or terns diving into a shoal of fish is memorable for any ornithologist. The birds steer their dives so that they have the maximum chance of catching a fish which they see beneath the surface. It seems likely that they benefit from fishing in groups, for a fish escaping from one diving tern is less likely to be able to avoid another tern diving immediately after the first. The birds sometimes benefit from the presence of large predatory fish for they drive the small fish to the surface where the birds can catch them. Collaboration by pelicans during fishing is described in the next Chapter.

Top left
The Kingfisher *(Alcedo atthis)* flies down from its perch and dives into the water in order to catch fish.

Left
A unique method of catching small fish, and other surface-living animals, is shown by the three species of skimmers *(Rhynchops)*. They are the size of a large tern and have a lower mandible which is considerably longer than the upper. This unusual bill is continuously dipped into the water as the bird flies along so that food items can be skimmed from the surface.

Right
The toucan's bill seems much less incongruous when the bird is seen using it to bite off large chunks of soft fruit. In the tropical forests frequented by toucans there is a tree species in fruit throughout much of the year.

Below
Hawfinches *(Coccothraustes coccothraustes)* have large, stout bills with which they can crack open large seeds. Birds which feed on small seeds, for instance redpolls of the genus *Acanthis,* have much smaller bills.

Above
The White-fronted Goose *(Anser albifrons)* uses its short broad bill for grazing. Geese are quite selective as to which plants they will eat, and will remove some plants from an area of short grassland while leaving others. They need to feed for much of the day in order to obtain enough digestible food, but they pass the food through their digestive system very rapidly so that they are never so overloaded with food that they cannot fly.

**Vegetarians**

The most abundant source of food on land is leafy vegetation. This is extensively eaten by insects and by some mammals, although very few birds feed directly on leaves. Much of the material in leaves is indigestible for birds, and even the insects and mammals need micro-organisms in their guts to help them to digest it. Those parts which can be digested by birds take longer to digest than does animal food. Since birds need to digest rapidly because they cannot carry large amounts of partially digested food when flying, most of them feed on foods with a greater concentration of available nutrients. Some birds, such as geese and pigeons, do eat green vegetation.

Seeds and fruit are much more readily digested by birds. Even in temperate countries where there are few specialized fruit-eaters many birds will take fruit when it is available. Insect-eaters such as warblers, and worm- or snail-eaters such as thrushes will eat blackberries and other small fruits when these are plentiful. In the tropics there are many fruit-bearing trees and one or more of these will be available at any time during the year. The tanagers and toucans are examples of specialist fruit-eaters. Seed-eaters also have specially adapted bills. They are heavier but smaller than those of fruit-eaters, for the bird needs to be able to exert considerable force at a particular point in order to be able to crack open the seeds.

Some birds feed on the secretions of plants. An example from temperate North America is the Yellow-bellied Sapsucker *(Sphyrapicus varius)* which makes holes in trees and drinks the sap. (If the sap has fermented the bird may become quite drunk!) Many tropical birds drink nectar as well as eating small insects. Many species of South American honeycreepers, Australian honeyeaters and African sunbirds feed in this way, but the most efficient nectar-eaters are the hummingbirds. The flowers which hummingbirds visit are usually pollinated by the birds, and not by insects. The shapes of these flowers and the shapes of the hummingbirds' bills have evolved together so that each is well adapted to the other. Hummingbirds with curved bills, such as the hermits *Glaucis* and *Phaethornis,* visit flowers with curved corolla tubes, and those with longer bills, such as starthroats *Heliomaster,* visit larger flowers than those with medium length bills such as the Ruby-throated Hummingbird *(Archilochus colubris).* This last species is a summer visitor to much of eastern North America.

The White-winged or Two-barred Crossbill *(Loxia leucoptera)* can extract seeds from pine and fir cones. The crossed tips of its upper and lower mandibles are inserted in the cone and the head is twisted to separate the scales.

# Courtship and nesting

## Social organization

All birds take part in some social interactions in additions to the minimum required for mating. Most have a true territory, that is to say a defended area, at some time in their lives and also show some parental care. The levels of complexity of such social behaviour and the extents to which there is aggregation are, however, very variable among birds.

Many species of birds, such as the Dunnock *(Prunella modularis)* from northern Europe, hold territory and interact with rivals, mate and reproduce during the breeding season but are largely solitary throughout the rest of the year. The Mute Swan *(Cygnus olor)* and various crows are examples of species which hold territory during the breeding season and then remain paired while moving around during the non-breeding period. Most geese and some game birds not only remain with their mates but allow their offspring to remain with them so that they travel in family parties at all times except during breeding. A further extension of such social behaviour is shown by Starlings *(Sturnus vulgaris)* and many other species which hold territory in pairs during the breeding season but spend some or all of the non-breeding period in groups which include many unrelated individuals.

Aggregations of birds into large groups may also occur during the breeding season. These aggregations may be brief associations called leks which persist for the period of courtship only or may involve social breeding. Social breeding with more or less social behaviour in the non-breeding period is shown by Gannets *(Sula bassana)* and various gulls, terns, auks and other seabirds. The most social birds which spend their whole lives close to other individuals are the babblers (Timaliinae) and various weaver birds (Ploceidae) and honeyeaters (Meliphagidae).

Aggregations of unrelated birds may occur while breeding, feeding or during the period when feeding is not possible. The most obvious aggregations during the non-feeding periods are the roosts which are formed at night by species which feed during the daytime. There are a few species of birds, such as the oilbird, which are like bats in that they group together in

This Black-headed Gull *(Larus ridibundus)* is adopting a threatening posture which would deter rivals from approaching.

caves during the daytime and feed at night, and there are many species which depend upon the shore for feeding and aggregate when the tide is in. The fact that not all species breed, feed, or roost in large groups leads to questions as to the functions for each species of continuous or intermittent communal living. It is noticeable that all of the most abundant species of birds live socially for some of the time.

Feeding flocks of Starlings *(Sturnus vulgaris)*; ducks and geese; waders on the shore; finches and sparrows; or seabirds, such as gulls and terns, are a familiar sight to ornithologists. The flocks may be denser when a localized concentration of food has been found but the flocks of those birds mentioned above usually remain together when moving from one feeding place to another and when resting. In suitable feeding areas, flocks may include hundreds or thousands of birds but there is much variation, even within species, in flock size. The most remarkable feeding flocks are those of the finch-like Quelea *(Quelea quelea)* of Africa which moves around in flocks numbering several millions. The communal roosts of many species are very impressive due to the presence of large numbers of individuals. Starlings roost on buildings in cities as well as in woodland. One of the most well-known roosts in Britain is that in Trafalgar Square in central London. Another much publicized roost in a city is the Pied Wagtail *(Motacilla alba)* roost in O'Connell Street, Dublin, Ireland.

Above
The Ruff *(Philomachus pugnax)* does not nest communally but the males form temporary gatherings called leks. Here they display their striking plumes and defend very small areas against other males. The females mate with the successful males.

Left
One of the best-studied European lek species is the Black Grouse *(Lyrurus tetrix)*; three males are shown displaying here. Leks are also formed by various other grouse species, hermit hummingbirds *(Phaethornis)* and the manakins (Pipridae) from South America.

The flight paths of Starling flocks into London have been traced easily because the passage of many thousands of birds over a city is very obvious. Another communal roosting species, which is conspicuous when flying to roost, is the Cattle Egret *(Bubulcus ibis)*, an Old World species which has been invading Central and North America during the last forty years. This species shares another famous roost with the Scarlet Ibis *(Eudocimus ruber)* in the Caroni Swamp, Trinidad.

Rivalling the large communal roosts as splendid ornithological sights are the vast breeding colonies of some seabirds. Some of the largest are of hole-nesters such as petrels and shearwaters but auks such as the Guillemot or Common Murre *(Uria aalge)* which nest on cliff ledges form some impressive breeding colonies. The pelicans, gannets, boobies and cormorants also breed in large conspicuous colonies, the most famous of which are the 'guano' islands off the west coast of South America. These colonies have been used by seabirds, feeding on the rich fauna of the coastal currents, for many centuries, and have become covered by a deep layer of bird droppings, which is the guano, much valued as a fertilizer in Peru. Some small birds build communal nests (see chapter on courtship and nesting) and many others are colonial breeders.

Above
The Gannet *(Sula bassana)*, which breeds socially on small islands with limited space for nests, builds its nest close to others, although not so close that the birds sitting on the nests can touch one another. A circle around the nest with radius equivalent to the neck plus bill length is the territory defended by the Gannet.

Left
Starlings *(Sturnus vulgaris)* roost communally in large numbers and travel to the roost along regular flight paths. Large flocks often perform impressive gyrations near the roost before settling for the night.

Occasionally, aggregations of birds may be accidental. Many birds may go to the same food source or to the same resting place but may derive no benefit from the presence of other members of their own species. In earlier discussions it was assumed that such accidents often explained the occurrence of large gatherings of birds, but when the distribution of potential food sources or resting places is considered it is usually obvious that the birds are choosing to go to a site because of the other birds which are there and not just because it is a suitable site.

One factor which is important to any small animal and will affect its distribution within its habitat is the avoidance of predation. Group living may help some species in this respect because the prey species may collaborate to fight off predators while individuals might not be able to do so. This is certainly an advantage for large birds such as crows or gulls, flocks of which can drive off most birds of prey. Smaller birds which have no chance of fighting off a predator might still derive some advantage from being in a flock, because predators may be unwilling to risk collisions which might occur if they flew into the middle of a dense flock. Alternatively, they might be confused by the large numbers and be unable to select an individual to attack. While feeding and roosting, groups of birds are much more likely than isolated birds to detect the approach of a predator and to give alarm calls which lead to the escape of all the members of the group. A division of labour is also possible in that some can watch for danger while others feed. These possible advantages of communal living are counteracted by the fact that the flock, roost, or colony is very much more obvious to predators than are individual birds or nests. Predators are often found in the vicinity of roosts and colonies and it is difficult to assess the extent to which the advantages of group living outweigh, or are outweighed by, the disadvantages when the avoidance of predation is considered.

The factor which has the greatest effect on bird behaviour throughout most of the year is food and its acquisition. The feeding movements of individual birds allow the efficient use of the specialized body form, bill shape, and so on, of the bird for obtaining food and the birds use optimum methods in hunting for food. It, therefore, seems likely that the widespread habit of spending time in social groups may confer some advantages in the exploitation of food resources. The most obvious advantage of feeding in a flock occurs when the food is concentrated more in some places than in others, rather than evenly distributed. If this is so, some individuals might find these local concentrations and inform the others in the flock with the result that all share the food. When a flock of tits is feeding in woodland they often call to one another and if one individual finds a particularly good tree for feeding, others soon arrive in the same tree. On another occasion, a different bird might find the food concentration so that it is to the advantage of all in the flock to call and to attract others in the flock to the food. A further advantage to the species is that young birds, which are not as good at finding food as are adults, will be shown food. If the food is evenly distributed, these advantages do not accrue from the flocking habit.

Another interesting result of birds living in groups is seen when a new food source is discovered. We do not normally see this occurring in field situations but there are several examples of birds starting to use new food sources provided by man. Soon after the use of silver-foil milk-bottle tops became widespread in Britain, a Great Tit *(Parus major)* was seen to tear a hole in one and to drink the cream on top of the milk. Soon afterwards other Great Tits in the area were seen to remove milk bottle tops in the same way and the habit spread to Blue Tits *(Parus caeruleus)*. It seems likely that each bird learned to remove bottle tops by watching other birds do it for, within a few years, tits in most parts of the country were removing bottle tops. A similar example was the spread of the habit of eating the berries of the garden shrub *Daphne* by Greenfinches *(Chloris chloris)*. Cultivation of this shrub became quite popular in Britain before any birds ate the berries. Then one year, Greenfinches started to eat the berries and the habit soon became regular in areas where the shrub was popular. Both of these new habits appeared suddenly and spread rapidly and both were shown by species which feed in flocks during the winter months. Experimental work has shown that young birds normally learn about the range of food sources available to them by copying other members of the flock.

Another advantage associated with flocking is available to birds feeding on small animals in earth or mud, for a large flock of birds is able to obtain the food in a more systematic manner. If Starlings *(Sturnis vulgaris)* searching for insect larvae moved randomly over a group of fields, birds would be spending much more time in areas already searched by other individuals. A more efficient way to

Left
Many birds which feed together during the day also roost communally. Bramblings *(Fringilla montifringilla)* are sometimes found together in large numbers during the winter in northern Europe.

Below
The White Pelican *(Pelecanus onocrotalus)* feeds in fresh water by scooping up fish with its enormous bill. When an individual scoops water, many fish escape by swimming away from the bill. The pelicans frequently feed in small groups which swim in a horseshoe pattern and then all move forward and scoop the water at the same time. Small fish which dart away from one pelican's bill are likely to be caught by another, thus all the birds benefit from feeding together.

search for the food is for flocks to move systematically across each field consuming the insect larvae as they go. Birds at the front of a moving flock walk around probing in the ground for the larvae. Those at the back who have exhausted the area in which they are searching, fly to the front and start searching in a new area. The overall effect is often a sort of rolling movement of the flock with the majority of birds searching on the ground and a few flying forwards. Sometimes Starlings in smaller flocks quarter the ground without needing to fly. The rolling movementof very large flocks is also seen in hunting Quelea *(Quelea quelea)* and in feeding locusts and it seems that this method of feeding in large flocks is the most efficient means of exploiting food resources of some kinds.

Birds which have to catch active prey may also benefit from feeding in flocks. Fish which escape from one bird in a flock may be caught by another. The White Pelican *(Pelecanus onocrotalus)* of east and South Africa and south-west Asia, which scoops fish from near the surface of the water while swimming; the skimmers which fly in V-formation, skimming fish and small crustaceans from the water; and terns diving for fish, all collaborate in this way.

It is apparent that flocking is worthwhile when food is distributed in certain ways but that widely spaced individual birds do best in other circumstances. Observations on two subspecies of *Motacilla alba,* White Wagtails in Israel and Pied Wagtails in England, have shown that this species feeds in flocks when there are local concentrations of food but may hold winter feeding territories when food is widely scattered. This is one of the species which forms communal roosts, and it is noticeable that species which do this feed for some or all of the time on food sources which are patchy in their distribution or which appear in a particular small area at one time and then disappear again. As a result of such observations it has been suggested that communal roosts may act as food information centres. When a group of birds finds a good food source on

one day, they are likely to fly directly to it on the next day. Thus, an individual which has previously been unsuccessful in finding food, but which spends the night in the roost, can follow groups which set off towards good feeding sites in the morning. Observations of departures from roosts of Quelea *(Quelea quelea)* and of Pied Wagtails show that post-roost gatherings are sometimes formed in prominent places by individuals arriving from the roost, and that birds depart in flocks from these gatherings towards feeding sites.

Some similar benefits could be obtained by birds which breed in colonies, for unsuccessful food finders could follow other individuals which had previously been more successful. Studies on colonial nesting Great Blue Herons *(Ardea herodias)* in British Columbia showed that the departure of one bird from the colony was frequently followed by that of others and that the herons usually joined groups of other herons feeding. These groups of herons were more successful in catching fish than were individuals feeding by themselves, probably because they did not need to spend as much of their time looking out for predators. The advantages for food finding which are conferred by group living must be balanced against possible disadvantages due to competition for food. Birds feeding in large numbers in close proximity to one another may collaborate in food finding but might sometimes survive better if more widely scattered. This competition will, however, tend to favour the best-adapted individuals so that they will survive better and the species will benefit.

There are several other possible advantages associated with communal living. It is easier to find a mate in such circumstances than if individuals are widely distributed. In a few species, packing close together in winter may conserve heat. For many species, complex population regulation mechanisms may operate more efficiently if the individuals spend some of their time concentrated in small areas. It is, however, difficult to decide which of these factors, or avoidance of predation, or efficiency of exploiting food resources, is the most to the evolution of the group-birds. It is even more difficult to decide which factor contributed most to the evolution of the group-living habit. The behaviour of birds in groups is adapted to minimize predation of birds in the group but this does not prove that defence against predators is the principal function of communal behaviour. Some food sources are probably exploited better by group-living birds but conclusive evidence for this is very difficult to obtain. It seems likely that some flocking and communal roosting arose initially because of the distribution of the food but it is quite possible that predation, or another factor, was important in initiating the habit in other species. For some species there are now several different advantages associated with group living so it is likely to persist, even in situations where it is temporarily disadvantageous.

## Territory and song

At the beginning of the breeding season, small birds living in mixed vegetation, such as the European Robin *(Erithacus rubecula)* or the east American Common Bluebird *(Sialia sialis)*, need to find a suitable nest site and need to be able to obtain adequate food for themselves and their young within easy flying distance from this nest site. One way of insuring that this food supply will continue to be available to the pair of birds which has built the nest is for that pair to drive away any other members of their own species which might compete with them for the food. The size of the territory, the area which is defended, will vary according to the size of the species and the density of its food supply. The tiny Least Flycatcher *(Empidonax minimus)* held territories of only 0·07 hectares (0·17 acres) in Michigan but Robins in England had territories of 0·6 hectares (1·48 acres) and the Western Meadowlark *(Sturnella neglecta)*, a species which lives in open country, defended 9·0 hectares (22·23 acres) against other members of its species. An extreme example is the Golden Eagle *(Aquila chrysaetos)* which had territories of 9,300 hectares (23,170 acres) in the mountains of California.

The male Bluethroat *(Luscinia svecica)* is typical of many small birds in that it proclaims the existence of its territory by singing for long periods during spring.

Above
The Little Tern or Least Tern *(Sterna albifrons)* has quite a large territory on a beach with some stones on it. This species shows much less collaboration in nest defence than do terns which nest in colonies, such as the Arctic Tern *(Sterna paradisea)* or Sandwich Tern *(Sterna sandvicensis)*.

Left
The Woodcock *(Scolopax rusticola)* flies around its territory in the evening making a call which starts as a low grunting and ends as a high whistle. This regular flight is sufficiently conspicuous to man to have been given a name—'roding'.

Since most wandering birds will refrain from entering an occupied territory, it is in the best interest of the owner to advertise the existence of the territory. A large species living in open country, such as an eagle, needs only to fly around high above the territory to make its presence obvious. The Wood Pigeon *(Columba palumbus)* demarcates its territory by flying around its perimeter with a characteristic series of upward flaps and downward glides. The most efficient way of advertising territory for most species is by the production of sounds. The roding Woodcock *(Scolopax rusticola)* calls as it flies and the Wood Pigeon often produces a clapping noise with its wings. Some species sing from a conspicuous song post within the territory where they can be seen as well as heard by rivals, but for many species, song from a concealed point is adequate advertisement.

The songs of some species are very elaborate but others are much simpler. In general, the more conspicuous the plumage of the bird, the less elaborate is its song. When a male bird arrives in its territory it often has to attract a mate as well as to repel potential rivals for the territory. In some species this is done principally by visual displays but in others, those with dowdier coloration, the song is more important (see next chapter on courtship and nesting). Some birds sing throughout much of the day when the territory is first proclaimed, for instance, most warblers (Sylviinae or Parulidae). Others, such as the Nightingale *(Luscinia megarhynchos)* in Europe or the Mockingbird *(Mimus polyglottus)* in North America, sing elaborate songs which they continue at night when most other species have stopped singing. The Mockingbird mimics other species so its mimicry is obvious to man but studies of various species suggest that an element of mimicry occurs in most songs. Each bird modifies its song as it develops by copying parts of the songs of other members of its own species which it hears singing while it is a juvenile and, later, of neighbours which hold territories near its own. This results in considerable geographical variation in song in some species.

The term territory always refers to a defended area but this varies greatly in size and function. The Robin's *(Erithacus rubecula)* territory is important in pair formation and serves as a familiar area which includes adequate food for the family of Robins. Its existence has some effect on breeding population density and its advertisement by display, including song, intensifies the pair bond as well as reducing interference and competition from rivals. In the Gannet *(Sula bassana)*, however, the territory only seems to keep rivals away from the nest itself and to identify the nest for its owners. Territories in leks are transient and serve only to form pairs and to allow selection of mates by females. In non-breeding periods, territories normally exist only when food is evenly distributed and are absent in many species.

Left
The male Robin *(Erithacus rubecula)* advertises its territory by singing, but should a rival male appear it attempts to frighten it away by fluffing out its red breast feathers.

Below
If a potential rival Robin fails to respond to song or display postures, as was the case with this stuffed bird, the territory owner will attack vigorously.

Above
Fights over territorial ownership are not often observed, but these Goldfinches *(Carduelis carduelis)* were obviously unable to settle their dispute by means of display.

# Flocks, roosts and territories

The Mandarin *(Aix galericulata)* drake has certain greatly expanded secondary feathers in its wing which stand up like orange flags on the back. These feathers are moved by the bill in a courtship display which is derived from preening movements.

## Display

Birds do not normally touch one another but preserve a certain minimum inter-individual distance. This distance may be smaller in species which live socially for much of their lives but it is still obvious among Starlings *(Sturnus vulgaris)* sitting in a row or feeding in a flock. If one individual comes too close to another it is actively repulsed. This behaviour is clearly beneficial to birds, for physical clashes between individuals might damage feathers and impair survival chances, but it raises problems when mating is necessary. In order that mating can occur, the male and female have to convince one another that no hostility is intended by their close approach. This is the principal reason why courtship displays are necessary. Once a display is performed by a species it can assume other functions. There is evidence that the displays of male doves must be detected by a female before the secretion of the hormones which lead to ovulation can occur. Ovulation is necessary before mating and successful fertilization is possible. The physiology of birds is linked to behaviour in a complex manner so that courtship and other displays are not just a prelude to mating but are also necessary for egg production, nest building, incubation and preservation of the pair bond.

The first step in reproduction is for the male and female to meet at the appropriate time of year and in suitable circumstances for pair formation to occur. Sometimes this is some weeks before nesting will be possible. In other species, they meet when the female enters a territory which has been established by the male. In most birds, the male is more brightly coloured, has the more elaborate song, or takes the initiative in display. The Dotterel *(Charadrius morinellus)* of the arctic and Old World mountainous areas, and the phalaropes are exceptional in that the females are larger and more brightly coloured than the males which incubate the eggs.

Among more normal species, females are often attracted by the same displays which are used to deter possible rivals. Males which hold adequate territories, or which are successful in leks or other competitive situations, are likely to be chosen by females, but those males which are not successful are ignored by females. Such selection has resulted in the evolution of many bizarre plumages and vocal specializations. Brightly coloured birds are at a disadvantage compared with duller birds because they are more obvious to predators. It has been suggested that the selection of brightly coloured males by females is advantageous to the species because a male, which can survive despite very conspicuous plumage, must be very well adapted in other respects and is therefore likely to be able to pass on useful genetic characters to its offspring.

One of the best-studied displays is that of the Great Crested Grebe *(Podiceps cristatus)* which has almost worldwide distribution except that it is replaced by the Western Grebe *(Aechmophorus occidentalis)* and related species in the Americas. The display involves a series of movements, which exaggerates the chestnut tufts on the grebe's head and is carried out by both members of the pair. One bird adopts a posture and then the other responds, often by adopting the same posture. One of the display movements involves diving to the bottom of the lake and emerging with a piece of weed in the bill while another involves both birds rearing up high out of the water and, in the Western Grebe, scooting rapidly over the

Top right
The most impressive part of the Peacock's *(Pavo cristatus)* display is the erection of the enormous fan of tail plumes. These are the upper tail coverts, which in most species are short feathers covering the base of the tail but in Peacocks are long with hair-like barbs except near the tip, where they are expanded to form large rounded 'ocellae' which have eye-like patterns on them.

Right
Calls form part of the displays of various North American grouse such as the Sage Grouse *(Centrocercus urophasianus)*. When calling, the male Sage Grouse expands the enormous greenish yellow air-sacs in its neck and makes a popping sound.

Right
The male Superb Lyrebird *(Menura superba)* from Australia looks somewhat like a game bird or bird of paradise, for it has a long decorative tail. During the elaborate display, the tail is erected and may then briefly resemble a lyre in shape. The bird also sings a loud song which includes imitations of various other birds.

Below
The frigate birds have a large gular sac just below the bill which is expanded like an enormous scarlet balloon when the male Great Frigate Bird *(Fregata minor)* displays.

surface together. Once the most intensive period of display is completed and the mating has occurred, the pair continue to indulge in displays sporadically throughout the time that they stay together.

Other groups of birds whose courtship displays have been intimately studied include gulls, grouse, some finches and ducks. The ducks are particularly interesting because it has been possible to discover the origins of some of the displays. Movements which now form part of courtship displays were often originally used in some other context but they have now become ritualized and are recognized by the partner. The Mallard *(Anas*

*platyrhynchos)* and several other related surface-feeding ducks include a movement in their display in which the head is turned and the bill is used to briefly move certain secondary feathers. This movement is also used during sequences of preening in which many other feathers are also moved but in the courtship context it is a brief flick of the bill and is therefore immediately distinguishable from a preening movement. It seems likely that this movement was occasionally carried out during the courtship displays of the ancestors of those ducks and gradually became incorporated in the display sequence. Movements like those used in preening have become incorporated in the displays of various other species but there are also movements with other origins. The Goldeneye *(Bucephala clangula)*, a duck found throughout most of the Northern Hemisphere, performs a display movement which involves throwing the head back and then forwards again. This is very similar to the movement of a Goldeneye which is about to take off from the surface of the water. Feeding movements, nest-building movements and movements shown by very young birds are among others which appear in some courtship displays.

The Peacock *(Pavo cristatus)* from India is one of the best-known examples of a species in which the male's plumage is very specialized in coloration and feather shape for courtship display. Several other members of the pheasant and grouse family have long tail plumes which are used in displays; both displays which attract females and drive off rivals and those which lead up to mating. The plumage adaptations of the birds of paradise are so attractive to man that a large trade in the feathers of these birds developed in the last century. Fortunately, most are now protected.

Air sacs and fleshy lobes, such as wattles and combs, are present in a variety of male birds and are used in display. Frigate birds and some grouse have large air sacs in the neck. Wattles are present in several cotingas (Cotingidae) and in all members of the New Zealand family of wattle birds (Callaeidae) as well as in the Domestic Fowl *(Gallus gallus)* and other game birds.

Another family which, like the birds of paradise, is commonest in New Guinea is the bower birds (Ptilonorhynchidae). Some of these are brightly coloured but most are dull and the display includes the use of objects gathered by the males. These objects range from a few leaves or petals in some species to an elaborate building many times the size of the bird in others. The 'bower' is brightest and largest in the species which are themselves the dullest in coloration. Bowers are built on the ground, usually of twigs and other vegetable material, and may be platforms, huts or avenues. A display area next to the construction is decorated by the

Right
Groups of male Greater Bird of Paradise *(Paradisea apoda)* display in trees and often hold display poses for some time while females look on. The feather adaptations of birds of paradise include long feathery plumes; expanded feathers which when erected together form helmet and beard-like shapes; wire-like feathers which are unadorned; and plate-like groups of barbs with a metallic appearance on otherwise wire-like feathers.

Below
Display between potential mates sometimes involves the offering or ritualized presentation of food. Each of these Black Guillemots *(Uria grylle)* is holding a sand eel *(Ammodytes)* and is posturing with it as part of the display.

Above
After elaborate courtship displays have reassured the members of a pair that each means the other no harm and have brought the female into the right physiological state, mating can occur. The species shown here is the Black-headed Gull *(Larus ridibundus)*.

Below
King Penguins *(Aptenodytes patagonica)* nest in large colonies called rookeries in Antarctica. They build no nest but incubate their eggs on top of their feet.

male bower bird with brightly coloured objects such as feathers, snail shells, flowers, fruits, stones, leaves, or man-made objects. Members of three genera paint their bowers using saliva mixed with fruit, grass pulp or wood. The Satin Bower Bird *(Ptilonorhynchus violaceus)* makes wads of fibrous bark and uses them as paint brushes to apply the coloured 'paint' to the bower. The bowers take a long time to make and they are always oriented in such a way that they will be optimally displayed when the morning sun strikes them. It is thought that the prolonged display period is necessary because, due to the vagaries of the season, the female may come into breeding condition and arrive at the bower at almost any time during the year.

## Nests

The great variety of nests produced by birds is due to the many different habitats which birds exploit and, as a consequence of this, to the different stages of development reached by birds before they hatch.

If a bird lives on the ground or on water and can escape from predators by running or swimming and hiding, it is necessary for its offspring to be able to accompany the parent as soon as possible after hatching. Young game birds, waders, ducks and rails are precocial. They hatch at an advanced stage of development so that they can walk and can feed by copying what their parents do. There is therefore no need for an elaborate nest. On the other hand, it is quite impossible for a bird to remain in its egg until it is well developed enough to be able to fly, so any bird which lives for most of its life off the ground must build a nest which will hold the young until they are able to fly. Once such a nest is available the bird can afford to be almost helpless at hatching. Birds like Robins *(Erithacus rubecula),* House Sparrows *(Passer domesticus)* and warblers, which can initially do no more than lift their bills to receive food, are called altricial.

Left
The Fairy Tern *(Gygis alba)* lays its single egg on a bare tree branch.

Below
The Sand Martin or Bank Swallow *(Riparia riparia)* is a colonial nester. Each pair excavates a nest hole in a sandy cliff.

Among precocial birds there are some which still have to build substantial nests because the eggs must be kept out of water. Flamin-

Left
House Martins *(Delichon urbica)* collect mud from the water's edge and use it to build a nest attached to a vertical wall.

Below
The nest of the tiny Ruby-throated Hummingbird *(Archilochus colubris)* is made of moss, spiders' webs and plant down. The parents' ability to hover means that the nest does not have to be designed so that the parent can land on it.

gos nesting in salt lakes build cones of mud with depressions in the top for the eggs. These keep the eggs clear of the water and also make it easier for the long-legged flamingo to sit on the eggs. A very elaborate nest is built by members of the Megapodiidae family such as the Mallee Fowl *(Leipoa ocellata)* of Australia or the brush turkeys *(Aepypodius* and *Tallegalla)* of New Guinea. The Mallee Fowl excavates a hole about 1 m (3·3 ft) deep and fills it with leaves and other vegetation which starts to decay. A hollow is made in the top of this and as many as twenty or thirty eggs are laid in it. These are then covered over with sandy soil to a depth of about 0·5 m (1·6 ft) and left to incubate in the heat produced by the decaying vegetation. The parents remain near the nest and test the temperature, removing or adding soil as required to maintain the temperature. When the eggs hatch, the young tunnel their way to the surface and immediately begin independent life. The brush turkeys, which live in dense jungle, build mounds which may be as large as 1 m (3·3 ft) high and 4 m (13 ft) across. In extreme contrast to these birds, penguins nesting on Antarctic ice build no nest and hatch their young on top of their feet.

One of the easiest ways for altricial birds to protect their eggs and helpless young is to utilize a

hole in a tree or a crevice in stones or earth. A further development among species which do this is to excavate a hole. Woodpeckers can drill holes in trees and other species such as nuthatches and tits can enlarge existing holes. Nuthatches can also reduce the size of apertures of holes in trees by the use of mud, and the remarkable nesting habits of hornbills, in which the male walls up the female in her tree-hole nest and feeds her through a small aperture, have already been mentioned in an earlier chapter. Sandy soil is excavated by the Sand Martin or Bank Swallow *(Riparia riparia)*, and by kingfishers, bee-eaters, puffins, burrowing owls and some parrots. Mud can be carried to a nest site and used as a cement in the construction of a nest. The House Martin *(Delichon urbica)* and Cliff Swallow *(Petrochelidon pyrrhonota)* make mud nests attached to vertical walls and the Ovenbird *(Furnarius leucopus)* of South America makes a hemispherical mud nest which may weigh 4 kg (8·8 lb). The Australian magpie larks *(Grallina)* also make large mud and grass nests, often in trees. Many other birds use some mud in the construction of their nests, for instance, the Chaffinch *(Fringilla coelebs)* and the Song Thrush *(Turdus philomelos)*

Reed Warblers *(Acrocephalus scirpaceus)* weave their nests around the stems of reeds at a height which renders them safe from rising water or predators reaching up from below, but not so high that they are blown around excessively by the wind.

which line their twig or fibrous nests with mud.

The most widespread nest-building technique is to weave twigs, grass or other vegetable material so as to form first a platform and then a cup-shaped structure. The smallest birds such as hummingbirds and tits use moss, lichen, spiders' webs and feathers

Right
The Striped Honeyeater *(Plectorhyncha lanceolata)* builds a delicately woven sac-like nest.

Below
The weavers are, as their name implies, among the most energetic of woven-nest builders. This Little Weaver *(Sitagra monacha)* builds a compact nest with the entrance on the side.

Top right
The massive communal nests built by the Sociable Weaver *(Philetairus socius)* in south-west Africa may be so large that they form a covering like a thatched roof over a whole tree with hundreds of funnel-mouthed nests underneath it.

Bottom right
The tailor birds sew leaves together with cobwebs. Downward pointing leaves are chosen and a series of knots is produced so that the top is open but the bottom halves of each of two leaves are sewn together. The nest cup thus formed is then lined and eggs are laid in it. This Long-tailed Tailor Bird *(Orthotomus sutorius)* chick is about to leave the nest.

but large birds such as crows and storks use small branches from trees. The nest is usually open at the top but there are many species which build a roof, for instance, Magpies *(Pica pica),* various tits, and American orioles which often build a pendulous nest with a side entrance. The largest nests are communal nests made by the many species of weaver and by Palmchats *(Dulus dominicus)* of Hispaniola in the West Indies. The Palmchat's nests are aggregations of single dwellings, each with its own entrance, in a multilayered structure. The birds collaborate in building the nest and several individuals are needed to carry the largest sticks in the nest for these may be almost 1 m (3·3 ft) in length.

# Eggs and young

There is considerable variation among birds in the number of eggs which they produce and in the amount of effort which the male and female birds devote to feeding and protecting their young. Breeding imposes a considerable strain on the birds for they must be able to find much more food in order to provide for themselves when defending territory, nest building, egg laying and feeding young, and they must provide adequate food for the young. Those species which eat food which has a very high nutritional value and which can be found in considerable quantity without too much expenditure of effort, can lay more eggs than those which must work harder for their food. It would be a waste of energy for a bird to lay eight eggs if it was incapable of rearing more than three young.

A further factor which will affect the amount of energy invested in the attempted production of offspring is the extent of predation. If two out of each three young birds of a species will be eaten by predators, the parents must produce more eggs and young than a species which is comparatively safe from predators. Birds which are not subject to much predation and which find adequate food without extremes of activity, such as some seabirds, can replace themselves by producing one young per pair every one or two years. If these birds laid more than one egg they

might not be able to feed the young for their food supplies may not include the seasonal peaks which those of land birds in temperate regions show. Some birds would gain no advantage by producing more young, even if they could rear them, for more would die during the following winter or dry season when there was inadequate food for the increased number of birds.

The behaviour of birds towards predators is considerably altered when they have young. The arrival of predators which might attack eggs or young will evoke alarm calls or defence when the adult would normally ignore such a predator. Terns usually just move a small distance away from people who walk near them but they will mob and dive at human intruders on nesting colonies. Anyone who has been struck on the head by the sharp bill of an Arctic Tern *(Sterna paradisea)* or perhaps by several individuals making a concerted attack, will avoid such colonies in future. The change in the behaviour of Mute Swans *(Cygnus olor)* during the breeding season is also noticed and heeded by most people.

Similar behaviour changes are shown towards other potential predators. Many plovers and related birds distract predators away from their nests or young by the 'broken-wing display'. The adult bird shuffles along the ground with a wing trailing in a very convincing manner and most predators respond by following it as it leads them away from the nest. Despite popular mythology, few adult birds will defend their nests to the death.

Above
Grebes, like this Great Crested Grebe *(Podiceps cristatus)*, build floating nests of twigs and, since the eggs would be vulnerable to predators flying overhead, they cover them with a few twigs or leaves before they depart from the nest.

Right
These Linnet *(Carduelis cannabina)* nestlings expose their brightly coloured gape when they beg. This is very conspicuous to the parents which are then stimulated to place food in it.

Adults have invested considerable effort in the production of their offspring and they behave in such a way as to protect that investment. Half-grown young represent a greater investment than recently laid eggs so they are normally defended more strongly. A live adult bird is, however, worth more in terms of reproductive potential than a nest with young for the adult can easily escape from most nest predators and breed again. The parent, therefore, has to decide how much damage to itself is tolerable in the defence of its young and to act accordingly.

Young birds are not passive in interactions with their parents. Even the least well-developed altricial birds are able to detect the arrival of the parent at the nest and to beg for food. As birds get older they become more specific in their begging. Very young Song Thrushes *(Turdus philomelos)* will beg when the nest vibrates, as it would when the parent lands upon it, but older thrushes will not beg unless an object of approximately the size of the parent with a body and a head is visible. Young Herring Gulls *(Larus argentatus)*, which are precocial, direct their begging towards the red spot on the lower mandible of the parents' yellow bill. Sandwich Tern *(Sterna sandvicensis)* chicks tend to aggregate in creches at the breeding colonies, and young birds respond to the advent of one of their parents as soon as they hear the parent bird call and do not respond to the calls of other birds. Many young precocial birds need to leave the nest and accompany their parents soon after hatching. They, therefore, show a following response as soon as they are able to walk. This response may be shown towards any large moving object and the characters of such moving objects which elicit the maximum following from young goslings, ducklings, domestic chicks and Moorhens *(Gallinula chloropus)* have been extensively studied. The young birds' preferences change with experience and a gosling which follows people during its first few days may subsequently show a preference for following people rather than members of its own species. In some species these preferences may continue until the bird is adult and, if it was reared by a foster parent of a different species, it may address its sexual displays to members of the foster parent species rather than its own. It must be very rare for any such problems to arise in the wild for young birds normally see their own parents much more than any other bird. Exceptions are brood parasites such as cuckoos or cowbirds. Many of these show subsequent preferences for the nest of the foster parent when they come to lay their own eggs but their sexual behaviour is addressed to their own species.

Below
The young ducklings of this Barrow's Goldeneye *(Bucephala islandica)* are following the parent as she swims along. The young birds will learn to feed by copying their parent.

Bottom
The begging behaviour of young Herring Gulls *(Larus argentatus)* is directed towards the conspicuous red spot on the parent's yellow bill. Experiments have shown that a plain yellow bill is less effective in eliciting begging. If a young bird is not selective in its begging, it may waste a lot of energy.

# Migration and navigation

The American Golden Plover *(Pluvialis dominica)* breeds in the high arctic, migrates down the western coasts of North and South America, winters in southern Brazil and Paraguay, and then migrates back north through Central America and the centre of North America.

For centuries, many people living in temperate countries have been aware of the fact that some birds are present in the summer but absent in the winter. The birds are not seen to arrive or depart and until this century much uncertainty existed about what happened to them. Even the perceptive observer Gilbert White, who wrote his *Natural History of Selborne* in 1789, was of the opinion that Swallows *(Hirundo rustica)* spent the winter at the bottom of ponds. The idea that many small birds might migrate to warmer countries for the winter months was not universally accepted until marked individuals had been proved to do so about sixty years ago. Before this time much evidence for migration had accumulated and most scientists were convinced that it must occur.

Swallows were abundant in Britain from April to September and in southern Africa for the remainder of the year. The spring and autumn movements of flocks of geese, cranes, pigeons and other larger birds were well known to shooters for they occurred during the daytime. Small birds were not seen migrating during the day but people noticed that great numbers sometimes swarmed around lighthouses and lightships on dark nights in spring and autumn and large influxes of small birds sometimes occurred at these times. Concentrations were often seen at peninsulas or islands near the coast or at gaps in mountain ranges and this led to the establishment of bird observatories at such places. The first of these was Heligoland off the north German coast and this was followed by others in Britain, Scandinavia and, later, in North America.

The arrival and departure times of relatively conspicuous birds like Swallows in northern Europe or bluebirds in North America were recorded during the last century and this provided some evidence for migration as the southern arrival dates were earlier than those further north. In the autumn, flocks of summer visitors were seen gathering and some people on coasts or in mountains saw flocks of small birds flying north in spring and south in autumn. Migrating birds were sometimes heard at night, just before they arrived in an area. A good example of this is the Redwing *(Turdus musicus)*, a thrush which flies south to Britain and other parts of western and southern Europe in the winter. The high-pitched flight call of this species can often be heard on cloudy nights in October and November as flocks fly in from Scandinavia or Russia.

The first conclusive evidence of migration came from bird-ringing studies. A pioneer scheme started in Denmark in 1899 and was soon followed by schemes in seven other European countries. North American schemes, where the word band is used instead of the word ring, were nationally organized by 1920. Each ring has the address of the ringing scheme and a number on it. The bird ringer completes a schedule for each set of rings recording the species, age and sex of each bird and the date and place of ringing. Anyone who finds a ringed bird and notifies the appropriate national authority, is informed of the previous history of that bird and the information is added to that accumulated for the species. On December 27th 1911, a Swallow which had been ringed in England as a nesting adult eighteen months earlier was caught in a farmhouse in Natal, South Africa. Since that time, hundreds more northern European Swallows have been caught in southern or central Africa. Almost as much excitement was aroused by the report of a Manx Shearwater *(Puffinus puffinus)* which was ringed as a nestling on Skokolm Island, Wales, in September 1960 and found dead in November 1961 on the south Australian coast. Since

Right
Migrating birds like these Starlings *(Sturnus vulgaris)* are often attracted to lighthouses and lightships at night. When the birds suddenly encounter fog or very overcast conditions, they may converge on lights in large numbers.

Below
During the early years of bird ringing or banding schemes, many of the birds ringed were pulli like this young Curlew *(Numenius arquata)*. The advent of mist nets has made it much easier to catch and ring fully fledged birds.

The Arctic Tern *(Sterna paradisea)* breeds in northern Europe, Asia and North America and then migrates to the Antarctic. Birds ringed in Britain have been recovered in South Africa and Australia as well as in the Antarctic Ocean. One ringed bird is known to have lived for twenty-seven years, during which time it must have flown about 800,000 km (500,000 ml) during migration alone.

that time, three ringed British terns, two Arctic and one Common, have been recovered in Australia. This sort of evidence is now available for most northern European birds and for many other migratory birds.

Other evidence about bird migration has been obtained in recent years by means of radar and, in a few studies, by the use of aeroplanes for following birds. Large birds and flocks of small birds can be detected on radar screens so that nocturnal migration can now be observed directly. The arrival of large numbers of migrants in an area has been found to coincide with waves of bird movement in the expected direction on the radar screens. Movements of birds to and from roosts and the ascent of swifts in the evening to spend the night high in the sky are also detected by radar. These images and those of migrating birds were a source of mystery to the people who operated the original radar sets during the Second World War. Images on radar screens which were not aeroplanes became known as angels.

The extent of migration differs according to climatic conditions. In central Canada almost all of the species which are fairly common at some time of the year spend the summer there and then migrate south while one or two species

arrive from further north during the winter. In northern Europe and the northern coastal regions of the USA, half to two-thirds of the birds are summer visitors; there are quite a few resident species and some winter visitors. In a tropical country, a high proportion of the birds are resident but a few migrants arrive during the northern and southern winters. The distances travelled also vary greatly among species. Small insectivorous birds, such as warblers, which breed in northern Europe may migrate to southern Europe or northern and central Africa. Various thrushes and finches which eat berries or seeds breed in Iceland, Scandinavia or western Russia and spend the winter in western and southern Europe. Many ducks such as the Wigeon *(Anas penelope)* breed in Russia and migrate to north-west Europe for the winter.

The principal advantages of migrating away from northern Europe or North America in winter is related to food supply. Birds which feed on active insects cannot find them at all in colder regions during the winter so they must migrate to survive. Some resident birds survive by changing their diet but the food supplies are much smaller in winter so there cannot be as many birds present. Food is abundant in the tropical and sub-

A Manx Shearwater *(Puffinus puffinus)* was the first species ringed in Britain to be recovered in Australia. Most of those breeding in northern Europe winter off the west coast of South America.

tropical winter quarters of most insectivorous birds but there are many birds and other animals competing for it. This competition and pressure from predators makes it more difficult for birds to breed in tropical conditions than in the summer of temperate regions. It is, therefore, worth the considerable effort of migrating long distances. The energy required for migration is provided by the accumulation of a fat reserve before departure. Birds weighed just before migration are often one-third heavier than the same healthy bird a few weeks earlier. The fat reserves are utilized during migration and birds which have just migrated have usually lost at least the extra one-third of their body weight. The Ruby-throated Hummingbird *(Archilochus colubris)* travels about 800 km (500 ml) across the Gulf of Mexico during migration. Early calculations of the amount of fat reserve built up by the birds and their energy consumption in hovering flight, suggested that the reserves were not sufficient for the journey. Subsequent studies have shown that more of the fat reserve is available for conversion into energy than had previously been thought, but we still do not know the rate of food reserve utilization during level flight. The fact remains that these tiny hummingbirds and many other birds are capable of remarkable feats of endurance during migration.

# Navigation

The mechanisms of bird navigation have long puzzled and intrigued ornithologists and the problems remain largely unsolved today. Migrant birds may travel many thousand kilometres to a wintering area and then return to the same breeding site the following year. Homing pigeons, and other birds, can be removed from their homes and released far away from it, yet they return safely. Various methods of navigation might be used by birds. It is certain that birds flying in a familiar area use conspicuous landmarks for navigation. Migrating birds will also follow guidelines such as sea coasts or mountain ranges. Such terrestrial cues must be important in many local movements of birds. They do not account for the return of pigeons from places which they have never visited before or the migration of birds over the sea, well out of sight of land. Experiments with Manx Shearwaters *(Puffinus puffinus)*, a species which is normally exclusively marine, showed that they could return overland to a nesting colony over 370 km (231 ml) away, at a speed which necessitated almost direct flight.

Information about direction of movement over the Earth's surface can also be obtained from the Sun, Moon and stars. Pigeon fanciers have long known that more birds get lost in cloudy conditions and birds have been observed to accumulate on coasts during a series of cloudy nights and then to disappear as soon as there is a clear night. The Sun's movement could be used to provide information about the direction of home. A bird which was displaced from home could measure the Sun's altitude and rate of change of altitude and relate this to the expected values for these measurements at home at the same time of day. In the Northern Hemisphere, if the Sun were too high, the bird would be too far south and should fly away from the Sun to go towards home. If the Sun were rising too steeply, the bird would be too far to the east and should fly in the approximate direction of the movement to compensate. Laboratory studies have shown that such measurements by birds are feasible and it seems likely that birds are able to use some such measurement of the Sun's movement to obtain at least a general compass bearing towards home. At night, the pattern of the major stars would have to be learned to obtain directional information. Studies on birds put in a planetarium, in which the apparent star pattern can be altered, show that adult migratory birds can use a stationary star pattern and adopt an appropriate direction of movement. Juvenile birds are not able to do this without training. When trained in a planetarium which had correct star movement patterns, young birds learned to orient using the Pole Star as the fixed point for north but, if the planetarium sky moved around the star Betelgeuse, the birds used this as their fixed point. The importance of an internal clock in some navigation is emphasized in experiments in which birds known to fly in a

particular direction were kept in captivity under a light-dark regime in which the clock was shifted by six hours. When these birds were tested, they flew or oriented in a direction which was 90° different from normal. Thus, a disruption of their clock by one-quarter of a day resulted in a change of their orientation of one-quarter of a circle.

The evidence for the use of celestial cues during bird navigation is extensive but some navigation is certainly possible when such cues are hidden and there are no familiar land marks. Radar studies have shown that birds can maintain orientation when flying between banks of cloud at night so that the stars would be invisible. Pigeons have been shown to home satisfactorily in completely overcast conditions, although such homing is usually less efficient than when the Sun is visible. The extreme example of homing using no visual cues was provided by an experiment in which pigeons were fitted with translucent covers over their eyes, which allowed them no patterned vision, and yet could still fly home. The mechanism which allows such homing is not known but it is disrupted by magnetic fields. Pigeons could home on sunny days even if small magnets were attached to their wings, so the navigation using the Sun is not affected by magnets. On overcast days, homing was possible for most normal birds but not for those with magnets on their wings. Laboratory studies have also shown that birds can orient in magnetic fields but it is not known whether this ability is sufficient to account for any of their homing. Recent studies have also indicated that pigeons could use their sense of smell for navigation in that they would fly upwind towards a home smell. Birds flying down coasts might maintain direction by hearing the low-pitched sound of the breakers on the shore.

In summary, it seems likely that birds have the ability to use visual landmarks, the Sun's movement, the star pattern, smell, hearing, or some other unknown sensory ability to navigate. The method used at any one time will be that which is most efficient in the particular circumstances.

Pigeon navigation has been extensively studied because pigeon racing is such a popular sport and pigeons are popular pets. Recent work with pigeons suggests that birds have at least three different methods of navigating.

# Economics and conservation

Below
The meat of the Domestic Fowl *(Gallus gallus)* is one of the cheapest varieties which farmers can produce. Feeding chickens and collecting their eggs is the most efficient of all the important methods of converting plant material into domestic animal protein. Both of these types of human food are likely to become more important as intensive poultry farming methods spread to more and more countries.

Right
The Guinea Fowl *(Numida meleagris)* is kept for food throughout much of the warmer parts of the world.

## Bird farming

The species of bird which is most important to man is the Domestic Fowl *(Gallus gallus)*. This must be the commonest bird for its estimated world population is over 3,000 million. The ancestor of the Domestic Fowl is generally agreed to be the Red Jungle Fowl of northern India, Assam, Thailand, Burma and Indo-China. The wild forms are rather smaller than commercial breeds. Domesticated fowl were known in the Indus valley and Persia before 2000 BC and they were used by all the great civilizations of the world including those of Central and South America, who must have obtained the first birds from Asia prior to the arrival of Columbus.

Other domesticated game birds include the Turkey *(Meleagris gallopava)*, the ancestor of which is the Mexican Wild Turkey; the Guinea Fowl *(Numida meleagris)* from Africa; the Peafowl *(Pavo cristatus)*; and various quails, pheasants and partridges. All are kept for their flesh and their eggs but some, especially the Peafowl, are also popular as ornamental birds. Among waterfowl kept for similar reasons are the Domestic Goose *(Anser anser)* which is descended from the Greylag Goose; the Chinese Goose *(Anser cygnoides)* the ancestor of which is the Swan Goose which breeds in Siberia and winters in China; the Domestic Duck *(Anas platyrhynchos)* which is the same species as the Mallard; and the Muscovy Duck *(Cairina moschata)* which was first domesticated from the wild swamp-dwelling form in South America. The most important characters which initially led to domestication must have been the quantity of flesh which could be obtained from individual birds and the ease with which the species became accustomed to man. Domestication of several species must have occurred more than once in different cultures. Some of these birds were initially kept for their eggs and others were regarded as sacred and kept for that reason. Breeders have subsequently selected those individuals which produced the most meat and eggs for the smallest amount of food. Other characters which have been considered in deciding upon which birds should be used for breeding have included their readiness to eat a variety of different sorts of food, docility, broodiness, fertility, resistance to diseases, body shape, egg colour and plumage.

## Pets

Most of the birds which have been domesticated for food have also been kept as pets and bred for their ornamental qualities. In addition to the Peacock, which is already very ornamental, breeds of fowl such as the silky, whose feathers are long and fluffy, and the Sebright bantam which has attractive white feathers with black edges, have been kept in various parts of the world. Like the Domestic Fowl, the first pigeons were domesticated more than four thousand years ago. All are descended from the Rock Dove *(Columba livia)* and many of the pigeons which are feral in towns have plumage which is quite similar to that of the wild form. This cannot be said of many domestic breeds for some are particularly bizzarre in plumage and in other characters. The fantails have monstrous tails and swollen necks so that the head rests on the base of the tail. Pouters have enormously swollen necks and fairy swallows have large fans of feathers around their legs like skirts. Carriers have thick bills and massive crinkled, fleshy ceres which overlap the bill and much of the front of the head. Some of these ornamental breeds particularly impressed Charles Darwin who used pigeons as examples in his work *On the Origin of Species* which was published in 1859. Pigeons have also been used for racing and for carrying messages. The message carrying has been of great use to man in certain circumstances. News of battles or other important events could be communicated to people near the pigeons' loft at about three or four times the speed of normal overland travel. Other examples of birds which have been made to work for man are the falcons which catch birds and mammals and the Chinese and Japanese Fishing Cormorants *(Phalacrocorax carbo)*. The fishermen keep the Fishing Cormorants on lines and put a ring around the base of the neck so that when the bird dives and catches a fish it can be pulled in and the fish removed by the fisherman.

Budgerigars *(Melopsittacus undulatus)* originate in Australia. They are very popular cage birds today, both for their coloration and for their ability as mimics.

Many other ornamental birds have been kept in cages by man. Some are also kept because of their songs. The number of species kept at some time is enormous but the most popular groups are the finches, parrots, doves and mynahs. The Canary *(Serinus canaria)* is a small finch which originated in the Canary, Madeira and Azores Islands in the Atlantic. It has been kept as a cage bird in Europe for at least 300 years and is particularly esteemed for its song. Canaries with plumage variations are also popular. Other finches which breed in captivity and which are popular cage birds are the Bengalese Finch *(Lonchura striata)* and Java Sparrow *(Padda oryzivora)* from Asia and the Zebra Finch *(Poephila castonotis)* from Australia. Regretably, there is still considerable trade in wild-caught finches, from India and other countries, for sale in Europe and North America as cage birds. No doubt the time will soon come when civilized people will refuse to keep birds and other animals as pets unless they have been bred in captivity.

Most birds which are kept in cages for their songs are passerines but some of the attraction of the smaller doves derives from their cooing songs. The particular ability of parrots, which has led to their popularity as cage birds for several centuries, is their mimicry of a variety of sounds including the human voice. The most popular species today is the Budgerigar *(Mellopsittacus undulatus)* from Australia. Some similar ability has been shown by members of the crow and starling families. The most proficient mimic of the human voice is the Hill Mynah *(Gracula religiosa)*. People have long been curious about the reasons why some birds mimic speech while others do not. Mynahs, like many birds, mimic their territorial neighbours in the wild. They can produce a wide range of sounds including those rather low-pitched sounds produced by man. Captive individuals which are kept with other mynahs will mimic them. Those kept with man produce sounds like the owners' voice or domestic noises such as telephones ringing, cutlery clinking or corks being withdrawn from bottles. When they produce such sounds, their owners usually respond by giving them increased attention which enlivens their rather boring lives.

## Bird pests

Man's domestic animals are either too large to be vulnerable to attack by predatory birds or, like most chickens, are kept indoors. A healthy lamb is too heavy for even a Golden Eagle to carry off. Crops, on the other hand, might be eaten by birds. By far the most important pests of crops are insects. The majority of birds eat insects so they do not compete with man. Insectivorous birds might be beneficial in a few circumstances but their impact on insect pest populations is generally too small to make any difference to the crop.

Without doubt, the most important bird pest is the Quelea or Red-billed Dioch *(Quelea quelea)* which lives in the savannah regions of Africa. Flocks of many thousands or even millions of birds move around the African continent feeding on the seeds of various members of the grass family. The average bird's diet is from 90 per cent to 100 per cent wild seed but occasional visits to fields of millet, guinea corn, or rice can result in immense damage. It is very easy to shoot them and a single shot may kill many birds. The reduction of numbers produced by this means is not sufficient to justify the expense of buying cartridges. Since Quelea spend the night in enormous communal roosts they can be attacked there. In Senegal in 1959, eighty million Quelea were killed by the use of explosive charges at roosts but the devastation of crops over the whole country by the birds was unaffected. Poisonous spray was used in South Africa in 1966 and 112 million birds were killed but again this made no difference to the amount of damage. There are several reasons for this: firstly, many millions of the birds would die each year anyway because only the fittest can survive; secondly, the population is very large, indeed, the Quelea may be the commonest bird in the world; thirdly, birds just move away from places where they are attacked to other parts of the country or to other countries; fourthly, the Quelea breed in one country and then, when conditions become adverse, they move to another country where they breed again. It has so far proved impossible to control total numbers in an economic way but effects on crops can be much reduced by frightening the birds away from intensive areas of crop production with loud noises made by people or by modern bird scarers. If crops are grown when natural food is abundant the damage is much less.

Problems with Quelea are similar to those with other agricultural bird pests. The Wood Pigeon *(Columba palumbus)* greatly reduces the production of green vegetable crops in some areas of Britain and, again, the best defence is to scare the birds away from the crops. There are several other bird pests of crops and a few minor pests of freshwater fisheries. Birds cause

Right
Hill Mynahs *(Gracula religiosa)* can mimic the human voice so well that their imitations of the voices of individual people are recognizable. It is noticeable that mynahs are most likely to produce a sound when people start to walk away from their cage, so that mimicry is a method of bringing people back to the mynah. One might say that mynahs train people to entertain them by rewarding them with sounds.

Below
Vast flocks of Quelea *(Quelea quelea)* move around the savannah lands of Africa feeding on seeds and occasionally causing enormous damage to man's crops. The governments of twenty-five countries support research into methods of controlling them. The bird is in the sparrow family but is smaller than a House Sparrow.

some serious damage in cities where House Sparrows and roosts of Starlings may foul buildings. A much greater problem is caused by birds being hit by aircraft. This problem has become much worse since the advent of jet engines since birds may be sucked into the engine and prevent its functioning. In most cases this merely damages the engine but occasionally a flock of birds is struck by a multi-engined plane and several engines are incapacitated at once, thus causing the plane to crash. Most of such incidents occur at take-off or landing so they can be reduced by scaring off birds which go near runways and by refraining from siting airfields near places where there are many birds. The damage to aircraft by birds is considerable. The Royal Air Force in Britain recently estimated that the annual cost to them of bird damage was about one million pounds sterling and the worldwide cost to civil and military aircraft may be one hundred times this amount. This figure and the much larger figure which is the result of damage to crops, seem large but they are very small in comparison with the damage caused by such animals as rodents, nematode worms or insect pests.

## Sport

The birds which are now domesticated on farms must originally have been hunted. Their relatives in the families of game birds, ducks, geese and various waders, pigeons, bustards, sandgrouse and others, are still eagerly hunted for food in various parts of the world. Even small birds are trapped and hunted in many countries. The contribution of such birds to the total animal protein consumption by man is extremely small and the food rewards of most hunting expeditions are slight. The main reason why the habit of hunting wild birds has persisted, even where the birds are not really needed for food, is simply that people enjoy it. The pleasure which can be obtained from this sport or from the rather more sophisticated sport of falconry, is quite readily comprehensible to most people today. The ancient sport of cock-fighting, however, is regarded with disgust in civilized countries. In some countries, such as Italy, the sport of hunting birds is generally accepted and, since most of the hunters have efficient guns, this results in a general scarcity birds. In other countries the pastime of watching birds has become so widespread that there are far more birdwatchers than hunters and the morality of hunting for pleasure is widely questioned. The first sign of this change in attitude in a country is normally an outcry against the killing of birds which have become scarce in that country. A pressure group forms which tries to persuade the government to pass laws protecting rare species. A further step is to prevent the killing of all birds except those regarded as pests or game. Presumably the next step will be the eradication of the game status.

Top
Bird reserves can be managed so that optimum conditions are maintained for a certain species. These islands at Minsmere in Suffolk, England are artificial and were constructed as nesting places for terns and feeding places for a variety of water birds.

Above
The public are encouraged to observe birds and other wildlife without disturbance, by the provision of hides or blinds at bird reserves.

Right
The Bald Eagle *(Haliaeetus leucocephalus)* is typical of species which are threatened by the accumulation in their bodies of toxic substances from their food. The species is now largely confined to areas of North America where agricultural and industrial chemicals are little used. There has been some recovery in their numbers in the last few years.

Above
The return of the Osprey *(Pandion haliaetus)* as a breeding bird in Scotland has been possible due to the presence of volunteer wardens at nest sites. Many tourists have seen the birds and shown interest in the conservation work.

# Conservation

The initial concern of those who enjoy seeing a wide variety of birds is to preserve the rarest species by preventing people from trapping them, shooting them, or taking their eggs. Such measures may prevent the local extinction of a species but if the population has dropped to a low level it may also be necessary to encourage the birds to stay in an area by providing improved feeding or nesting places. Examples of this in Britain are the successful efforts of the Royal Society for the Protection of Birds to encourage the nesting of Avocets *(Recurvirostra avosetta)* in Suffolk and Ospreys *(Pandion haliaetus)* in Scotland. Similar operations on a worldwide scale have been launched by the World Wildlife Fund.

In order to preserve rare species it is first necessary to know about their requirements so research into their behaviour, physiology, and general ecology is encouraged. Efforts may also be made to increase populations of rapidly declining species by the introduction of the species from elsewhere or by breeding the birds in captivity. The introduction of species has led to many problems in the past so there is a general reticence to transport birds from one country to another. Breeding in captivity has certainly resulted in previously rare species being helped back to a viable wild population. One of the first examples was the Hawaiian Goose *(Branta sandvicensis)* whose world population had declined to fewer than fifty in 1950. Breeding programmes in Hawaii and, especially, at the Wildfowl Trust, Slimbridge, England resulted in many hundreds of individuals being returned to the wild. Important breeding programmes are at present in operation in Maryland, USA for rare species such as the Whooping Crane *(Grus americana)* and the Bald Eagle *(Haliaeetus leucocephalus)*.

A different approach to conservation, which is becoming more generally recognized as the most valuable one, is to preserve types of habitat rather than individual species. Since species have continually arisen and disappeared during animal evolution, it can be argued that preservation, in some cases, is not desirable. A more important attitude is to prevent the loss or despoiling of whole habitats due to man's activities. Marshland is generally unproductive since it cannot be used for agriculture or for building at present so it has has gradually been disappearing from all countries. This can be remedied by keeping areas as reserves for plants and animals. This is not very costly and the preservation of examples of all types of habitat is likely to bring benefits in food production and environmental control when ecology is more thoroughly studied. The other conservation problem is to prevent all habitats from being drastically altered by substances produced by man such as the persistent agricultural or industrial chemicals. Man is part of the fauna of Earth and changes which are brought about by man's activities are not necessarily undesirable because they are due to man rather than to any other species. The fact remains that we have some control over the rate and extent of change so we must consider our priorities and our obligations to future generations in deciding what changes to allow.

Above
The Wood Pigeon *(Columba palumbus)* is the only bird which is a serious pest of agriculture in Britain. Their success is due in part to their ability to produce 'pigeon milk' in their crops. This substance is fed to the young and allows them to develop rapidly without the necessity for the parents to change their diet. Despite being large birds, they can therefore produce several broods in one breeding season.

Right
The establishment of bird reserves by the Royal Society for the Protection of Birds in areas in England where Avocets *(Recurvirostra avosetta)* were intermittent breeders resulted in regular breeding by the species.

# Index

*Page numbers in italic refer to illustrations.*